JACARANDA
GEOGRAPHY ALIVE 9

AUSTRALIAN CURRICULUM | SECOND EDITION

COORDINATING AUTHOR

JILL PRICE

CONTRIBUTING AUTHORS

JILL PRICE

CATHY BEDSON

JANE WILSON

ELYSE CHORA

KINGSLEY HEAD

MARILYN WIBER

DENISE MILES

ALISTAIR PURSER

CLEO WESTHORPE

KERRY BAINBRIDGE

PAT BEESON

jacaranda
A Wiley Brand

Second edition published 2018 by
John Wiley & Sons Australia, Ltd
42 McDougall Street, Milton, Qld 4064

First edition published 2013

Typeset in 11/14 pt Times LT Std Roman

© John Wiley & Sons Australia, Ltd 2018

The moral rights of the authors have been asserted.

National Library of Australia
Cataloguing-in-publication data

Author:	Price, Jill, author.
Title:	Jacaranda Geography Alive 9 for the Australian Curriculum / Jill Price, Cathy Bedson, Jane Wilson, Elyse Chora, Kingsley Head, Marilyn Wiber, Denise Miles, Alistair Purser, Cleo Westhorpe, Kerry Bainbridge, Pat Beeson
Edition:	Second edition.
ISBN:	978-0-730-34713-2 (paperback)
Notes:	Includes index.
Target Audience:	For secondary school age.
Subjects:	Geography — Textbooks. Geography — Study and teaching (Secondary) — Australia.
Other Creators/ Contributors:	Bedson, Cathy, author. Wilson, Jane, author. Chora, Elyse, author. Head, Kingsley, author. Wiber, Marilyn, author. Miles, Denise, author. Purser, Alistair, author. Westhorpe, Cleo, author. Bainbridge, Kerry, author. Beeson, Pat, author.

Trademarks
Jacaranda, the JacPLUS logo, the learnON, assessON and
studyON logos, Wiley and the Wiley logo, and any related trade
dress are trademarks or registered trademarks of John Wiley &
Sons Inc. and/or its affiliates in the United States, Australia and in
other countries, and may not be used without written permission.
All other trademarks are the property of their respective owners.

Front cover image: In Green / Shutterstock

Cartography by Spatial Vision, Melbourne and MAPgraphics
Pty Ltd, Brisbane

Illustrated by various artists and the Wiley Art Studio.

Typeset in India by diacriTech

This textbook contains images of Indigenous people who are, or
may be, deceased. The publisher appreciates that this inclusion
may distress some Indigenous communities. These images
have been included so that the young multicultural audience for
this book can better appreciate specific aspects of Indigenous
history and experience.

It is recommended that teachers should first preview
resources on Indigenous topics in relation to their suitability for
the class level or situation. It is also suggested that Indigenous
parents or community members be invited to help assess the
resources to be shown to Indigenous children. At all times the
guidelines laid down by the relevant educational authorities
should be followed.

Printed in Singapore

M WEP162343 270922

CONTENTS

How to use the *Jacaranda Geography Alive* resource suite...vii
Acknowledgements xi

1 The world of Geography   1

1.1 Overview...1
1.2 Geographical concepts 3
1.3 Review 11

UNIT 1   BIOMES AND FOOD SECURITY   13

2 All the world is a biome   15

2.1 Overview...15
2.2 SkillBuilder: Describing spatial relationships in thematic maps   `online only`
2.3 What is a biome?...17
2.4 How do we use the grassland biome?..20
2.5 What are coastal wetlands?...22
2.6 SkillBuilder: Constructing and describing a transect on a topographic map   `online only`
2.7 Why are coral reefs unique?..25
2.8 What are Australia's major biomes?...28
2.9 Why are biomes different?...30
2.10 How do we protect biomes?   `online only`
2.11 Review   `online only`

3 How can we feed the world?   36

3.1 Overview...36
3.2 How can we feed the world?..37
3.3 What does the world eat? 40
3.4 How does traditional agriculture produce food?...43
3.5 How did Indigenous Australian peoples achieve food security?...46
3.6 SkillBuilder: Constructing ternary graphs   `online only`
3.7 How have we increased our food?...50
3.8 How are biomes modified for agriculture? 53
3.9 How is food produced in Australia? 55
3.10 What does a farming area look like? 59
3.11 SkillBuilder: Describing patterns and correlations on a topographic map   `online only`
3.12 Why is rice an important food crop? 62
3.13 Why is cacao a special food crop?   `online only`
3.14 Daly River: a sustainable ecosystem?   `online only`
3.15 How can aquaculture improve food security for Indigenous Australian peoples?   `online only`
3.16 Review   `online only`

4 What are the impacts of feeding our world?   67

4.1 Overview 67

4.2 How does producing food affect biomes? 68

4.3 Where have all the trees gone? 70

4.4 SkillBuilder: GIS — deconstructing a map .... online only

4.5 Paper profits, global losses? .... online only

4.6 Should we farm fish? 74

4.7 SkillBuilder: Interpreting a geographical cartoon .... online only

4.8 How do we lose land? 80

4.9 Irrigation: success or failure? 82

4.10 Does farming use too much water? .... online only

4.11 Why is global biodiversity diminishing? 84

4.12 Does farming cause global warming? .... online only

4.13 Review .... online only

5 Are we devouring our future?   88

5.1 Overview 88

5.2 Who's not hungry? 89

5.3 Who is hungry? 92

5.4 SkillBuilder: Constructing and describing complex choropleth maps .... online only

5.5 How does a famine develop? .... online only

5.6 Will land loss lead to food shortages? 95

5.7 SkillBuilder: Interpreting satellite images to show change over time .... online only

5.8 Are we running dry? 100

5.9 Climate change: freeze or fry? 103

5.10 Why is food being wasted? .... online only

5.11 Review .... online only

6 2050 — food shortage or surplus?   107

6.1 Overview 107

6.2 Can we feed the future world population? 108

6.3 Can we improve food production? 111

6.4 What food aid occurs at a global scale? 115

6.5 SkillBuilder: Constructing a box scattergram .... online only

6.6 Do Australians need food aid? 118

6.7 Is trade fair? 121

6.8 SkillBuilder: Constructing and describing proportional circles on maps .... online only

6.9 How do dietary changes affect food supply? 124

6.10 Can urban farms feed people? 126

6.11 Review .... online only

7 Geographical inquiry: Biomes and food security   130

7.1 Overview 130

7.2 Process 131

7.3 Review 132

8 How do we connect with places? 135

8.1 Overview 135
8.2 How do we 'see' places? 136
8.3 SkillBuilder: Interpreting topological maps .... online only
8.4 How do we move around our spaces? .... online only
8.5 What does our land mean to us? 139
8.6 How do places change? 142
8.7 How do we access places? 144
8.8 To walk or not to walk? 149
8.9 SkillBuilder: Constructing and describing isoline maps .... online only
8.10 Are we all on an equal footing? 153
8.11 How do we connect with the world? .... online only
8.12 Review .... online only

9 Tourists on the move 158

9.1 Overview 158
9.2 What is tourism? 159
9.3 Who goes where? 164
9.4 SkillBuilder: Constructing and describing a doughnut chart ... online only
9.5 Who comes and goes in Australia? 167
9.6 SkillBuilder: Creating a survey .... online only
9.7 SkillBuilder: Describing divergence graphs .... online only
9.8 What are the impacts of tourism? 171
9.9 What can we learn from our travels? 174
9.10 Are zoos and aquariums eco-friendly? .... online only
9.11 What is cultural tourism? 176
9.12 Is sport a new tourist destination? 179
9.13 Review .... online only

10 Buy, swap, sell and give 183

10.1 Overview 183
10.2 How does trade connect us? 184
10.3 How does trade connect Australia with the world? 187
10.4 How is food traded around the world? 191
10.5 SkillBuilder: Constructing multiple line and cumulative line graphs ... online only
10.6 How has the international automotive trade changed? 195
10.7 Are global players altering the industrial landscape? 198
10.8 Why is fair trade important? 200
10.9 Why does Australia give foreign aid? 203
10.10 Why is the illegal wildlife trade a cause for concern? .... online only
10.11 SkillBuilder: Constructing and describing a flow map .... online only
10.12 Review .... online only

11 For better or worse? 208

11.1 Overview 208

11.2 How do you communicate? 209

11.3 Who has access to technology? 212

11.4 What are the consequences of unequal access? 214

11.5 How has technology improved lives in developing countries? 217

11.6 What are the impacts of e-waste production and consumption in China? 219

11.7 How are e-wastes managed? 223

11.8 SkillBuilder: Constructing a table of data for a GIS .... online only

11.9 How does e-cycling work? .... online only

11.10 How can you reduce your consumption? 228

11.11 SkillBuilder: Using advanced survey techniques — interviews .... online only

11.12 Review .... online only

12 Fieldwork inquiry: What are the effects of travel in the local community? 232

12.1 Overview 232

12.2 Process 233

12.3 Review 234

Glossary 235

Index 239

HOW TO USE

the *Jacaranda Geography Alive* resource suite

For more effective learning, the *Jacaranda Geography Alive* series is now available on the learnON platform. The features described here show how you can use *Jacaranda Geography Alive* to optimise your learning experience.

'Geographical concepts' is a valuable reference section that covers each of the seven concepts.

Each concept is clearly defined.

A variety of visual resources support the explanations.

A series of activities to build and develop your understanding of each concept is provided.

Activities provide you with an opportunity to apply all of the seven concepts.

Linking to *myWorld Atlas* will deepen your understanding.

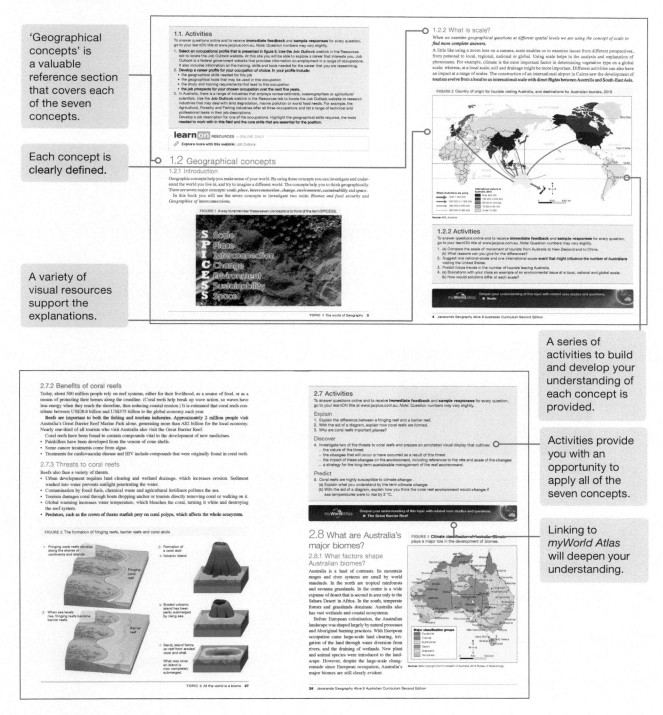

TOPIC 2
All the world is a biome

2.1 Overview

Numerous **videos** and **interactivities** are embedded just where you need them, at the point of learning, in your learnON title at www.jacplus.com.au. They will help you to learn the content and concepts covered in this topic.

2.1.1 Introduction

Where do the foods we eat and the natural products we use daily come from? Biomes. Biomes are communities of plants and animals that extend over large areas. Some are dense forests; some are deserts; some are grasslands, like much of Australia; and so the variations continue. Within each biome, plants and animals have similar adaptations that allow them to survive.

Biomes can be terrestrial (land based) or aquatic (water based). Understanding the diversity within them is essential to our survival and wellbeing.

Within each biome, there are many variations in the landscape and climate, and in the plants and animals that have adapted to survive there.

Starter questions
1. As a class, develop your own definition of the word biome.
2. What information about biomes is conveyed by the images shown on the previous page?
3. Brainstorm a list of biomes.
4. Select one of the biomes you listed in question 3, and create a mind map of the unique features you think it has.
5. What type of biome do you live in? If you cannot decide, why do you think that is?
6. Make a list of the things you use that have come from a biome. Compare your list with other students in your class.

INQUIRY SEQUENCE
2.1	Overview	15
2.2	SkillBuilder: Describing spatial relationships in thematic maps	16
2.3	What is a biome?	17
2.4	How do we use the grassland biome?	20
2.5	What are coastal wetlands?	22
2.6	SkillBuilder: Constructing and describing a transect on a topographic map	25
2.7	Why are coral reefs unique?	25
2.8	What is Australia's major biomes?	28
2.9	Why are biomes different?	30
2.10	How do we protect biomes?	35
2.11	Review	35

2.2 SkillBuilder: Describing spatial relationships in thematic maps

WHAT ARE SPATIAL RELATIONSHIPS IN THEMATIC MAPS?

A spatial relationship is the interconnection between two or more pieces of information in a thematic map, and the degree to which they influence each other's distribution in space. Describing these relationships helps us understand how one thing affects another.

Go online to access:
• a clear step-by-step explanation to help you master the skill
• a model of what you are aiming for
• a checklist of key aspects of the skill
• a series of questions to help you apply the skill and to check your understanding.

FIGURE 1 Thematic map of Asia showing biomes

learnON RESOURCES — ONLINE ONLY

Watch this eLesson: Describing spatial relationships in thematic maps (eles-1726)

Try out this interactivity: Describing spatial relationships in thematic maps (int-3344)

Questions raise issues, link the unit to your life, and prompt you to think about what you already know and feel about the unit.

A sequence for your inquiry.

A thought-provoking topic opener sets the scene for your inquiry.

Evocative and informative images stimulate interest and discussion.

Each section begins with a clearly identifiable subtopic number and inquiry question.

Aquatic biomes

Water covers about three-quarters of the Earth and can be classified as fresh water or marine. Freshwater biomes contain very little salt and are found on land; these include lakes, rivers and wetlands. Marine biomes are the saltwater regions of the Earth and include oceans, coral reefs and estuaries. Marine environments are teeming with plant and animal life, and are a major food source. Elements taken from the roots of mangroves have been used in the development of cancer remedies. Compounds from other marine life have also been used in cosmetics and toothpaste.

FIGURE 6 Water

2.3 Activities

To answer questions online and to receive **immediate feedback** and **sample responses** for every question, go to your learnON title at www.jacplus.com.au. Note: Question numbers may vary slightly.

Remember
1. Name the five major biomes of the Earth and classify them as either aquatic or terrestrial.
2. Identify the broad characteristics biomes share.

Explain
3. Look carefully at figure 1. Using geographic terminology and concepts (including reference to latitude), describe the location and characteristics of the five major biomes.
4. (a) The map in figure 1 identifies five basic biomes. However, the text tells you that within each there are variations. Suggest reasons for these variations.
   (b) Are all biomes equally important? Explain your answer.

Predict
5. Select one of the categories of biomes described in this section. Suggest how this biome might be *changed* and used by humans and what impact this *change* might have on the environment.

learnON RESOURCES — ONLINE ONLY

Try out this interactivity: Beautiful biomes (int-3317)

2.4 How do we use the grassland biome?

2.4.1 What are the characteristics of grasslands?

Grassland, pampas, savanna, chaparral, cerrado, prairie, rangeland and steppe all refer to a landscape that is dominated by grass. Once, grasslands occupied about 42 per cent of the Earth's land surface, but today they make up about 25 per cent of its land area. Grasslands are found on every continent except Antarctica.

The grassland biome is dominated by grasses, and generally has few or no trees, though there may be more tree cover in adjoining areas, such as along riverbanks. They develop in places where there is not enough rain to support a forest but too much rain for a desert; for this reason they are sometimes referred to as a transitional landscape.

Grasslands are found in both temperate and tropical areas where rainfall is between 250 mm and 900 mm per year. In tropical regions, grasslands tend to have a distinct wet and dry season. In temperate regions, the summers tend to be hot and the winters cool. Generally, grasslands in the southern hemisphere receive more rainfall.

Grasslands can occur naturally or as a result of human activity. The presence of large numbers of grazing animals and frequent fires prevent the growth of tree seedlings and promote the spread of grasses. Unlike other plant species, grasses can continue to grow even when they are continually grazed by animals, because their growth points are low and close to the soil. Because grasses are fast-growing plants, they can support a high density of grazing animals, and they regenerate quickly after fire.

Some grasses can be up to two metres in height, with roots extending up to one metre below the soil.

FIGURE 1 Grasslands occupy about a quarter of the Earth's land surface.

World grasslands

Source: Spatial Vision

Easily identifiable visual material is referenced in the text and in activities.

Italicised key concepts are applied to the activities.

A wide range of engaging and informative visuals are included.

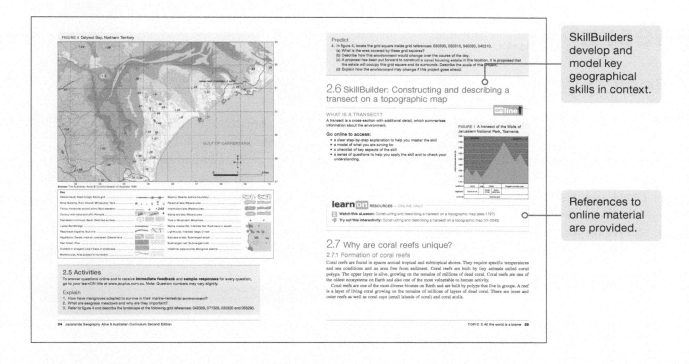

SkillBuilders develop and model key geographical skills in context.

References to online material are provided.

The Fieldwork inquiry and Geographical inquiry provide you with an opportunity to develop your inquiry skills in the field and through research.

UNIT 2 GEOGRAPHIES OF INTERCONNECTIONS

TOPIC 12
Fieldwork inquiry: What are the effects of travel in the local community?

12.1 Overview

Numerous **videos** and **interactivities** are embedded just where you need them, at the point of learning, in your learnON title at www.jacplus.com.au. They will help you to learn the content and concepts covered in this topic.

12.1.1 Scenario and your task

People travel for many reasons at the local scale — for example, they may travel to work, to shops, to visit friends and to local sporting venues. Often there are times when traffic congestion occurs, creating danger areas for motorists and pedestrians. Examples of places where such congestion occurs are schools and shopping centres. Undertaking fieldwork allows you to observe and collect original data first-hand.

Your task

Your team has been commissioned by the local council to compile a report evaluating the impacts of travel movements around a local school or traffic hotspot. You will need to collect, process and analyse suitable data and then devise a plan to better manage future traffic and pedestrian movement in the area.

12.2 Process

12.2.1 Process

- As part of a class discussion, determine a suitable location for your fieldwork study. This might be your own or a local school, or a nearby shopping centre. Talk about some of the issues related to your fieldwork site and then devise a key inquiry question — for example: What are the effects of … ? or How can we reduce the impact of … ? This will be the focus of your fieldwork. You then need to establish the following:
  - **What** sort of data and information will you need to study the travel issue at your site?
  - **How** will you collect this information?
  - **Where** would be the best locations to obtain data?
  - **When** would be the best times of the day or day(s) of the week to obtain data?
  - **How** will you record the information you are collecting?
  If you wish to collect people's views on the issue, or suggestions for improvements, you will need to plan and write suitable survey questions.

12.2.2 Collecting and recording your information and data

- As a class, plan the field trip by identifying and allocating tasks and possible sites to groups or pairs. It is often easier to share data collection. Once everything has been planned, you will need to perform your allocated tasks on the day.
- In class, invite your school principal or a member of your local council to be a guest speaker discussing your fieldwork site. They may be able to assist with background information that you may not be able to gain elsewhere. They can also provide a different perception of the effects of travel at your site. Plan a series of questions you would like to ask and be prepared to take notes that you can use in your report.
- After the field trip, it may be necessary to collate everyone's data and summarise surveys so that everyone has access to the shared information.

12.2.3 Analysing your information and data

- Look at your completed graphs and maps. What trends, patterns and relationships can you see emerging? Within your fieldwork area, are there some places that have a bigger issue with cars and pedestrians than other areas?

  Is there an interconnection between traffic congestion and time of the day, or day of the week? What have your surveys revealed? What are the major effects of travel at your fieldwork site? How do people perceive the travel issues in this place? Go back to your key inquiry question. To what extent have you been able to answer it? Write your observations up as a fieldwork report using subheadings such as:
  - Background and key inquiry question
  - Conducting the fieldwork [planning and collecting data]
  - Findings [results of data analysis].
- Download the report template from the Resources tab to help you complete this project. Use the report template to create your report.

Inside your *Jacaranda Geography Alive learnON*

Jacaranda Geography Alive learnON is an immersive digital learning platform that enables real-time learning through peer-to-peer connections, complete visibility and immediate feedback. It includes:

- a wide variety of embedded videos and interactivities to engage the learner and bring ideas to life
- the **Capabilities** of the Australian Curriculum, available in and throughout the course in activities and **Discussion** widgets
- links to the *myWorld Atlas* for media-rich case studies
- sample responses and immediate feedback for every question
- **SkillBuilders** that present a step-by-step approach to each skill, where each skill is defined and its importance clearly explained
- collaborative activities
- and much more.

ACKNOWLEDGEMENTS

The authors and publisher would like to thank the following copyright holders, organisations and individuals for their assistance and for permission to reproduce copyright material in this book.

Images

• AAP Newswire: **141**/Dan Peled • AEGIC: **192** (bottom)/Australian Bureau of Statistics • AIHW: **154**/Source: Australian Institute of Health and Welfare. Licence at https://creativecommons.org/licenses/by/3.0/au/. • Alamy Australia Pty Ltd: **2** (top)/SCPhotos; **8**, **128** (top)/Dinodia Photos; **44** (bottom)/© Tor Eigeland; **45** (bottom)/Jim Zuckerman; **57** (middle)/© AGF Srl; **63** (middle)/© Nigel Cattlin; **65**/Radius Images; **85** (middle).d/AfriPics.com; **96** (bottom)/© Frans Lanting Studio; **110** (top)/© Kim Haughton; **135**/Hilke Maunder; **158**/© T.M.O.Travel; **181** (bottom)/© Elliot Nichol; **208**/© Andrew McConnell; **217**/© Benedicte Desrus • AMTA - MobileMuster: **229** (top)/MobileMuster - The official recycling program of the mobile phone industry • Andrew J. Haysom: **143** (top).b • Australian Bureau of Statistics: **119** (middle left)/Based on data from: Australian Bureau of Statistics 2015, Australia Age Structure in 2015: Estimated Resident Population, ABS, Canberra.; **119** (middle right)/Source: Australian Bureau of Statistics. Licence at https://creativecommons.org/licenses/by/2.5/au/.; **209**/ Source: Australian Bureau of Statistics 2008, ABS Household Use of Information Technology, Australia, 2008-09 cat. no. 8146.0, ABS, Canberra.; **212**/Australian Bureau of Statistics 2009, 4901.0 - Childrens Participation in Cultural and Leisure Activities, Australia, Apr 2009, ABS, Canberra. • Australian Made Campaign Ltd.: **198** • Bike Paths & Rail Trails: **147**/Map courtesy the Bike Paths & Rail Trail Guide Victoria www.bikepaths.com.au • City of Ballarat: **155** • Copyright Clearance Center: **78** (top)/OECD/Food and Agriculture Organization of the United Nations 2015, OECD-FAO Agricultural Outlook 2015, OECD Publishing, Paris. http://dx.doi.org/10.1787/agr_outlook-2015-en • Creative Commons: **24** (top)/The Australian Army © Commonwealth of Australia 1999; **143** (bottom).c/Wikimedia Commons / Chris Brown • David Beaumont: **47** • denisbin: **49**/Flickr • Department of Agriculture,: **166** (bottom)/DAFF 2012, National Food Plan green paper 2012, Department of Agriculture, Fisheries and Forestry, Canberra. CC BY 3.0; **195** (middle right)/DAFF 2013, Australian food statistics 2011–12, Department of Agriculture, Fisheries and Forestry, Canberra. CC BY 3.0 • Department of Education and Training: **189** • Department of Environment, Land, Water & Planning : **150** (top)/State of Victoria • Department of Foreign Affairs and Trade: **188**, **189** (top), **190**/Department of Foreign Affairs and Trade website — www.dfat.gov.au. Licence available at https://creativecommons.org/licenses/by/3.0/au/.; **193**, **205**/Department of Foreign Affairs and Trade website – www.dfat.gov.au • Dept of Sustainability & Env.: **229** (bottom)/© Commonwealth of Australia 2013. • ECF Farmsystems: **127** • Fairtrade Australia: **121** (bottom) • Fairtrade South Africa: **202** (top) • FAO: **38** (bottom)/Source: Food and Agriculture Organization of the United Nations, 2015, World agriculture:towards 2015/2030 - Summary report, Table: Crop yields in developing countries, 1961 to 2030, http://www.fao.org/docrep/004/y3557e/y3557e08.htm#l, 13/05/2016, Reprodu; **74**/ OECD-FAO Agricultural Outlook; **81**/Source: Food and Agriculture Organization of the United Nations, 2011, The State of the World's Land and Water Resources for Food and Agriculture Summary Report, http://www.fao.org/nr/water/docs/ SOLAW_EX_SUMM_WEB_EN.pdf, Reproduced with permission.; **93** (middle)/Source: Food and Agriculture Organization of the United Nations, 2015, The State of Food Insecurity in the World , http://www.fao.org/3/a-i4646e.pdf, Reproduced with permission.; **96** (top)/Food and Agriculture Organization of the United Nations, Reproduced with permission; **113**/ Source: Food and Agriculture Organization of the United Nations, YIELD GAP FOR A COMBINATION OF MAJOR CROPS, http://www.fao.org/fileadmin/templates/solaw/images_maps/map_6.pdf, Reproduced with permission.; **124**; **195** (middle left)/c FAO 2012, Food and Agriculture Organization of the United Nations, http://faostat3.fao.org/home/index. html. This is an adaptation of an original work by FAO. Views and opinions expressed in the adaptation are the sole responsibility of the author or aut • FAPRI-ISU: **125**/Dermot Hayes • Featherbrook: **152**, **152**/MUST BE PRINTED BELOW IMAGE: Central Equity, Featherbrook Point Cook • Geoscience Australia: **10**, **60**/© Commonwealth of Australia Geoscience Australia 2017. Licence at https://creativecommons.org/licenses/by/4.0/.; **61**/© Commonwealth of Australia Geoscience Australia 2013. With the exception of the Commonwealth Coat of Arms and where otherwise noted, this product is provided under a Creative Commons Attribution 3.0 Australia Licence. • Getty Images Australia: **41** (bottom)/DEA / S. VANNINI; **44** (top)/Nicole Duplaix; **52**/© Ingetje Tadros; **64** (bottom)/UIG; **224**/Alvin J. Baez-Hernandez/ASAblanca • Global Harvest Initiative: **109** (top)/Re-drawn from an image by Global Harvest Initiative 2011 GAP Report®: Measuring Global Agricultural Productivity, data from the United Nations. • International Journal of Advanced Research in Artificial Intelligence : **218**/Oteri, Omae Malack; Kibet, Langat Philip; Ndung'u Edward N.; TABLE III: Mobile Subscriptions, 1999 to 2013, International Journal of Advanced Research in Artificial Intelligence, Vol. 4, No.1, 2015 • IPCC: **103** (bottom)/ Figure 3.6 from Climate Change 2007: Synthesis Report. Contribution of Working Groups I, II and III to the Fourth Assessment Report of the Intergovernmental Panel on Climate Change [Core Writing Team, Pachauri, R.K. and Reisinger, A. eds.]. IPCC, Geneva • ISAAA: **114** (top)/James, C. 2016. Global Status of Commercialized Biotech/GM Crops: 2016, ISAAA Brief 52, Ithaca, NY, USA. http://www.isaaa.org • John Wiley & Sons Australia: **9**, **75** (top) • MAPgraphics:

45 (top), **123** (bottom) • Maplecroft: **91**, **92**/Verisk Maplecroft - Verisk Maplecroft's Food Security Index provides a quantitative assessment of risks to the continued availability, stability and access to sufficient food supplies. The index also considers the nutritional outcomes of each countrys rel • Meals on Wheels NSW: **120** • NASA Earth Observatory: **12** (top left), **12** (top right), **67** (a), **67** (b), **71**, **97** • Newspix: **232**/Jono Searle • Out of Copyright: **48** (top)/National Library of Australia • Oxford University Press U.K: **43**/Waugh, D. 2000. Geography: An Integrated Approach, 3rd edition. Surrey: Oxford University Press, figure 16.25, p. 478. Reproduced by permission of Oxford University Press. • Panos Pictures: **39** (top)/Sven Torfinn/CABI; **199**/Fernando Moleres • Science and Information SAI: **218** (right)/Figure 5: A map of mobile coverage, Omae Malack Oteri, Langat Philip Kibet and Ndung'u Edward N., "Mobile Subscription, Penetration and Coverage Trends in Kenya's Telecommunication Sector" International Journal of Advanced Research in Artificial Intellige • SecondBite: **119** • Shutterstock: **1**/Scott Prokop; **2** (middle).a/Dmitri Ma; **2** (middle).b/Dmitry Kalinovsky; **2** (middle).c/ Toa55; **2** (bottom).a/avemario; **2** (bottom).b/JB Manning; **2** (bottom).c/eakkachai; **3** (bottom)/Christian Draghici; **5**/hans engbers; **17**/Joseph Sohm; **17** (bottom)/Evgeniya Moroz; **18** (top)/Kaesler Media; **18** (middle)/© Eric Isselee; **18** (bottom)/© Nicram Sabod; **19**/© Vladimir Melnikov; **21** (top right)/THPStock; **21** (bottom)/© gillmar; **29** (top)/Marco Saracco; **29** (middle)/© Richard Whitcombe; **29** (middle)/totajla; **29** (bottom)/kuehdi; **30** (top)/© Janelle Lugge; **31** (middle)/© Gbuglok; **34** (top)/© Snaprender; **36**/Em7; **41** (top)/© Hurst Photo; **42**/© CHEN WS; **53** (bottom)/© Sebastian Radu; **57** (bottom)/© Orientaly; **58** (top)/© Rosamund Parkinson; **58** (middle)/© Kaesler Media; **59** (bottom)/Phillip Minnis; **62** (bottom)/Zzvet; **63** (top)/© John Bill; **75** (bottom)/© Andreas Altenburger; **76** (top)/© Anneka; **78**/© Sukpaiboonwat; **80**/ Dirk Ercken; **82**/© Phillip Minnis; **85** (top).a/© Oleg Znamenskiy; **85** (middle).b/© John Wollwerth; **85** (middle).c/© Bye-likova Oksana; **85** (bottom)/© Michel Piccaya; **88**/R.M. Nunes; **90** (top)/I. Pilon; **90** (middle)/Hector Conesa; **93**/© paul prescott; **100** (bottom)/Pataporn Kuanui; **107**/CRSHELARE; **111**/Federico Rostagno; **130**/Robyn Mackenzie; **131**/nito; **131** (top)/Natalya Rozhkova; **132**/wavebreakmedia; **136**/Zurijeta; **143** (top).a/TK Kurikawa; **143** (bottom).d/Javen; **145** (top)/ Kummeleon; **145** (bottom left)/taewafeel; **145** (bottom right)/Nils Versemann; **146**/Stephen Finn; **150** (bottom)/boreala; **156**/ Richard Thornton; **160** (bottom)/AVAVA; **162** (top)/© iralu; **163** (bottom)/Pretty Vectors; **164** (bottom)/Lucky Business; **177**/123Nelson; **178**/plavevski; **180** (top)/Anthony Ricci; **183**/cozyta; **187** (top).d/© nikkytok; **187** (middle).a, **187** (middle).c/© Pressmaster; **187** (middle).b/© bikeriderlondon; **187** (bottom).e/© Goodluz; **187** (bottom).f/© Levent Konuk; **189**/Robert Kneschke; **190**/Claudine Van Massenhove; **198**/testing; **201** (top)/Mohammad Saiful Islam; **211**/ bloomua; **223**/asharkyu; **226**/rezachka; **232** (top)/Russell Shively; **234**/Laszlo66 • Spatial Vision: **4**/Australian Bureau of Statistics 2015, Overseas Arrivals and Departures, Australia, Jun 2015, cat. no. 3401.0, ABS, Canberra. & Website of Tourism Research Australia. Map drawn by Spatial Vision.; **6**, **225**/Data from Greenpeace. Map drawn by Spatial Vision.; **7**, **105** (top)/Data from Reducing climate change impacts on agriculture: Global and regional effects of mitigation, 2000–2080 by Tubiello F N, Fisher G in Technological Forecasting and Social Change 2007, 747: 1030-56. Map drawn by Spatial Vision.; **20**, **54**, **83** (top), **193**, **196**, **197**, **199**; **28** (bottom)/Data copyright Commonwealth of Australia, 2013 Bureau of Meteorology. Map drawn by Spatial Vision.; **38** (top)/Data courtesy of the Institute on the Environment IonE, University of Minnesota. Map drawn by Spatial Vision; **53** (top)/Source: American Geophysical Union and Google Maps. Image created by Spatial Vision.; **56**/© Commonwealth of Australia Geoscience Australia 2013. Map drawn by Spatial Vision.; **69** (bottom)/ Data courtesy of the Institute on the Environment IonE, University of Minnesota. Map drawn by Spatial Vision.; **77**/Data from PEW Environment Group. Map drawn by Spatial Vision.; **86**/© Commonwealth of Australia Department of Sustainability, Environment, Water, Population and Communities 2013. Map drawn by Spatial Vision.; **98** (top)/Data from GRAIN, 2008. Map drawn by Spatial Vision.; **98** (bottom)/Data from Friends of the Earth. Map drawn by Spatial Vision.; **101** (bottom)/Data from the Centre for Environmental Systems Research, University of Kassel. Map drawn by Spatial Vision.; **105**/Data from the European Commission. Map drawn by Spatial Vision.; **122**/Data from Fairtrade Foundation. Map drawn by Spatial Vision.; **143** (middle)/© OpenStreetMap contributors. Map drawn by Spatial Vision.; **145**/Data from Australian Bureau of Statistics Year Book Australia, 2012, cat. no. 1301.0, ABS, Canberra & United Nations Development Programme. Map drawn by Spatial Vision.; **160**, **165** (top), **206**; **168** (top); **169**/Data © Commonwealth of Australia Geoscience Australia 2013 & © State of Queensland Department of Agriculture, Fisheries and Forestry 2013. Map drawn by Spatial Vision.; **184**/Data from Wikmedia Commons. Map drawn by Spatial Vision.; **192** (top)/Data from World Trade Organization. Map by Spatial Vision.; **213** (top)/Data from The World Bank. Map drawn by Spatial Vision. • StEP Initiative United Nations: **219**, **220**, **221**, **222**/E-Waste in China: A Country Report • Sundrop Farms Pty Ltd: **114** • Telegeography: **210** • The Citizen: **148**/Daryl Holland • The World Bank: **185**, **186**/The World Bank: Household final consumption expenditure per capita, World Development Indicators; **213** (bottom)/World Bank. 2016. The Little Data Book on Information and Communication Technology 2015. Washington, DC: World Bank. doi:10.1596/978-1-4648-0558-5. License: Creative Commons Attribution CC BY 3.0 IGO Map drawn by Spatial Vision.; **215**/The World Bank: Internet users per 100 people • Turqle Trading: **201** (bottom) • UNDP: **101** (middle)/2006 Human Development Report, United Nations Development Programme. http://hdr.undp.org/en/content/copyright-and-terms-use • UNTWO: **161** (middle left)/© UNWTO, 92844/39/17. World Tourism Organization 2011, Tourism Trends, Assessment and a Glimpse of UNWTO", presentation by Xu Jing, Director, Regional Programme for Asia and the Pacific, UNWTO, during the Training Programme on Tourism Marketing, Tianjin • UNWTO: **159** (bottom)/World Tourism Organization 2013, UNWTO Tourism Highlights, 2015 Edition, UNWTO,

Madrid, p. 5.; **161** (middle right)/World Tourism Organization 2013, UNWTO Tourism Highlights, 2015 Edition, UNWTO, Madrid, p. 6.; **165** (middle)/© UNWTO, 92844/41/17. World Tourism Organization 2013, <i>UNWTO Tourism Highlights, 2013 Edition</i>, UNWTO, Madrid, p. 13. • Vanessa Harris: **137** • Viscopy: **140**/© Donkeyman Lee Tjupurrula/Licensed by Viscopy, 2016 • Wikimedia Commons: **23** (middle), **26** (middle).b/NOAA Photo Library; **26** (middle).c/Derek Keats; **48** (bottom)/CSIRO; **181** (top)/Nick J Webb; **195**/Rich Niewiroski Jr. / Creative Commons 2.5; **204** (bottom)/Based on: Marcos Elias de Oliveira Júnior, Data from Human Development Report 2015. • World Food Programme: **116** • World Trade Press: **218** (left)/© Copyright 2015 by World Trade Press. All Rights Reserved. • WorldAtlas.com: **62** (top)/WorldAtlas.com, http://www.worldatlas.com/articles/the-countries-producing-the-most-rice-in-the-world.html • ziilch: **228**/Reproduced with permission from ziilch

Text

• © Australian Curriculum, Assessment and Reporting Authority (**ACARA**) 2010 to present, unless otherwise indicated. This material was downloaded from the Australian Curriculum website (www.australiancurriculum.edu.au) (**Website**) (accessed September 5, 2017) and was not modified. The material is licensed under CC BY 4.0 (https://creativecommons. org/licenses/by/4.0). Version updates are tracked on the 'Curriculum version history' page (www.australiancurriculum. edu.au/Home/CurriculumHistory) of the Australian Curriculum website. • ABC: **198** • Department of Foreign Affairs and Trade: **188**/Department of Foreign Affairs and Trade website – www.dfat.gov.au • FAO: **51** (bottom)/Source: Food and Agriculture Organization of the United Nations, 2015, World agriculture:towards 2015/2030 - Summary report, http://www.fao.org/nr/water/docs/SOLAW_EX_SUMM_WEB_EN.pdf, Reproduced with permission. • International Food Policy Research Institute: **51** (top)/Source: Bumb and Baanante 1996. Reproduced with permission from the International Food Policy Research Institute. • MKG Group Database: **165** (bottom) • Redfin: **151** (top)/Walk Score®, Transit Score®, Bike Score™, ChoiceMaps™ • The World Bank: **215** (left), **215** (right)/World Bank. 2016. The Little Data Book on Information and Communication Technology 2015. Washington, DC: World Bank. doi:10.1596/978-1-4648-0558-5. License: Creative Commons Attribution CC BY 3.0 IGO • WorldAtlas.com: **41** (middle)/World Atlas, http://www.worldatlas.com/articles/world-leaders-in-corn-maize-production-by-country.html

Every effort has been made to trace the ownership of copyright material. Information that will enable the publisher to rectify any error or omission in subsequent reprints will be welcome. In such cases, please contact the Permissions Section of John Wiley & Sons Australia, Ltd.

TOPIC 1
The world of Geography

1.1 Overview

Numerous **videos** and **interactivities** are embedded just where you need them, at the point of learning, in your learnON title at www.jacplus.com.au. They will help you to learn the content and concepts covered in this topic.

1.1.1 Work and careers in Geography

Geographical skills will be useful for your future employment. An understanding of Geography and its application for managing sustainable futures is pivotal knowledge that will be desirable to future employers. In Geography, students are developing an understanding of the world. The skills you develop in Geography are transferrable to the workplace and can be used as a basis for evaluating strategies for the sustainable use and management of the world's resources.

Skills for work

Geography is a foundational skill for many occupations. Learning to navigate further education and training paths will help you to understand the variety of occupations that the study of Geography can lead to. The study of Geography includes important geospatial and spatial technology skills. These skills underpin the knowledge base of a range of courses and careers. Start your pathways exploration by considering who might use the key geospatial and spatial technologies.

- *Geospatial skills*: the ability to collect and collate information gathered from fieldwork and observations. Geospatial skills are used in careers such as surveying, meteorology, agricultural science and urban planning.
- *Spatial technologies*: technologies that demonstrate the connections between location, people and activities in digital formats. Jobs in the spatial industry are varied and include working in business and government. Spatial technologies apply many techniques, such as photogrammetry, remote sensing and global positioning systems (GPS). Spatial technologies manage information about the environment, transportation and other utility systems.

FIGURE 1 GIS (geographical information systems) being used to manage spaces and plan escape routes during a fire

FIGURE 2 Using GPS to survey and record road traffic movements for a local council

1.1.2 Where can Geography lead?

There is a range of careers that utilise Geography as a foundation skill. As you consider your pathway options for senior studies you may like to research some of the careers that are provided in figure 3.

FIGURE 3 Geography pathways

Meteorologist	Surveyor	Landscape architect
Meteorologists use geographical skills to forecast the weather, study the atmosphere and understand climate change.	Surveyors use geographical skills to measure, analyse and report on land-related information for planning and development.	Landscape architects use geographical skills to plan and design land areas for large-scale projects such as housing estates, schools, hospitals, parks and gardens.

Agricultural technician	Park ranger	Environmental manager
Agricultural technicians use geographical skills to support and advise farmers on aspects of agriculture such as crop yield, farming methods, production and marketing.	Park rangers use geographical skills to support and maintain ecosystems in national parks, scenic areas, historic sites, nature reserves and other recreational areas.	Environmental managers use geographical skills for project management and the development of environmental reports.

1.1. Activities

To answer questions online and to receive **immediate feedback** and **sample responses** for every question, go to your learnON title at www.jacplus.com.au. *Note*: Question numbers may vary slightly.

1. Select an occupational profile that is presented in figure 3. Use the **Job Outlook** weblink in the Resources tab to locate the Job Outlook website. At this site you will be able to explore a career that interests you. Job Outlook is a federal government website that provides information on employment in a range of occupations. It also includes information on the training, skills and tools needed for the career that you are researching.
2. Develop a career profile for your occupation of choice. In your profile include:
   - the geographical skills needed for this job
   - the geographical tools that may be used in this occupation
   - the study and training requirements that lead to this occupation
   - the job prospects for your chosen occupation over the next five years.
3. In Australia, there is a range of industries that employs *conservationists, oceanographers* or *agricultural scientists*. Use the **Job Outlook** weblink in the Resources tab to locate the Job Outlook website to research industries that may deal with land degradation, marine pollution or world food needs. For example, the Agricultural, Forestry and Fishing industries offer all three occupations and list a range of technical and professional tasks in their job descriptions.
   Develop a job description for one of the occupations. Highlight the geographical skills required, the tools needed to work with in this field and the core skills that are essential for the position.

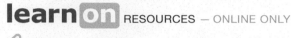
1.2 Geographical concepts

1.2.1 Introduction

Geographic concepts help you make sense of your world. By using these concepts you can investigate and understand the world you live in, and try to imagine a different world. The concepts help you to think geographically. There are seven major concepts: *scale*, *place*, *interconnection*, *change*, *environment*, *sustainability* and *space*.

In this book you will use the seven concepts to investigate two units: *Biomes and food security* and *Geographies of interconnections*.

FIGURE 1 A way to remember these seven concepts is to think of the term SPICESS.

1.2.2 What is scale?

When we examine geographical questions at different spatial levels we are using the concept of scale to find more complete answers.

A little like using a zoom lens on a camera, scale enables us to examine issues from different perspectives, from personal to local, regional, national or global. Using scale helps in the analysis and explanation of phenomena. For example, climate is the most important factor in determining vegetation type on a global scale; whereas, at a local scale, soil and drainage might be more important. Different activities can also have an impact at a range of scales. The construction of an international airport in Cairns saw the development of tourism evolve from a local to an international scale with direct flights between Australia and South-East Asia.

FIGURE 2 Country of origin for tourists visiting Australia, and destinations for Australian tourists, 2015

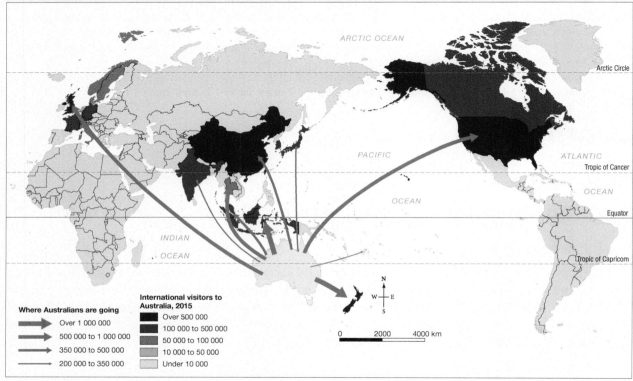

Source: ABS, Austrade

1.2.2 Activities

To answer questions online and to receive **immediate feedback** and **sample responses** for every question, go to your learnON title at www.jacplus.com.au. *Note*: Question numbers may vary slightly.

1. (a) Compare the *scale* of movement of tourists from Australia to New Zealand and to China.
   (b) What reasons can you give for the differences?
2. Suggest one national-*scale* and one international-*scale* event that might influence the number of Australians visiting the United States.
3. Predict future trends in the number of tourists leaving Australia.
4. (a) Brainstorm with your class an example of an environmental issue at a local, national and global *scale*.
   (b) How would solutions differ at each *scale*?

 **Deepen your understanding of this topic with related case studies and questions.**
⊙ **Scale**

1.2.3 What is place?

The world is made up of places, so to understand our world we need to understand its places by studying their variety, how they influence our lives and how we create and change them.

Everywhere is a place. Each of the world's biomes — for example, a desert environment — can be considered a place, and within each biome there are different places, such as the Sahara Desert. There can be natural places — an oasis is a good example — or man-made places such as Las Vegas. Places can have different functions and activities — for example, Canberra has a focus as an administration centre, while the MCG is a place for major sporting events and the Great Barrier Reef is a place of great natural beauty with a coral reef biome. People are interconnected to places and people in a wide variety of ways — for example, when we move between places or connect electronically via computers. We are connected to the places that we live in or know well, such as our neighbourhood or favourite holiday destination.

FIGURE 3 Large-scale farming of green peppers in a greenhouse where soils, moisture, nutrients and the weather are all controlled

1.2.3 Activities

To answer questions online and to receive **immediate feedback** and **sample responses** for every question, go to your learnON title at www.jacplus.com.au. *Note*: Question numbers may vary slightly.

Refer to figure 3.
1. Why do you think people have *changed* this place by building greenhouses there?
2. What characteristics of a desert biome are being altered in this *place*?
3. What features might this location have for the production of food?
4. What would be the advantages and disadvantages of greenhouse farming?
5. Suggest the types of crops that would be suitable for greenhouse farming.
6. List ways in which people living in other *places* in Europe may be *interconnected* to the greenhouses in Almeria.

 **my World** Atlas **Deepen your understanding of this topic with related case studies and questions.**
○ **Place**

1.2.4 What is interconnection?

People and things are connected to other people and things in their own and other places, and understanding these connections helps us to understand how and why places are changing.

Individual geographical features can be interconnected — for example, the climate within a place or biome, such as a tropical rainforest, can influence natural vegetation, while removal of this vegetation can affect climate. People can be interconnected to other people and other places via employment, communications, sporting events or culturally. The manufacturing of a product may create interconnections between suppliers, manufacturers, retailers and consumers.

FIGURE 4 Many countries have strict laws dealing with e-waste disposal. It is often easier to export the material to countries in South-East Asia, where there are fewer laws and the wastes can be broken down, recycled or sold. It is extremely hazardous.

Source: Spatial Vision

1.2.4 Activities

To answer questions online and to receive **immediate feedback** and **sample responses** for every question, go to your learnON title at www.jacplus.com.au. *Note:* Question numbers may vary slightly.

Refer to figure 4.
1. Describe the distribution of the main ports dealing with e-waste trade.
2. What is the *interconnection* between the main recycling countries and the main ports that receive and dispatch e-waste?

3. Where are the destinations of e-wastes from Europe?
4. Describe the *interconnection* between North America and Asia in relation to e-waste trade.
5. Create a diagram to show the *interconnections* that could occur for the growing, manufacturing, sales and consumption of a can of pineapple slices.

 Deepen your understanding of this topic with related case studies and questions.
❯ Interconnection

1.2.5 What is change?

The concept of change is about using time to better understand a place, an environment, a spatial pattern or a geographical problem.

From a geographical time perspective, change can be very slow — think of processes such as the formation of mountains or soil. On the other hand, a volcanic eruption or landslide can change landforms rapidly. It may take some years for the boundary of a city to expand outwards, but in the space of a few weeks whole suburbs can be demolished to make way for a freeway. Change can also have physical, economic and social implications. Consider the effect of the internet over the past few years.

FIGURE 5 Predictions of the effects of climate change on cereal crops

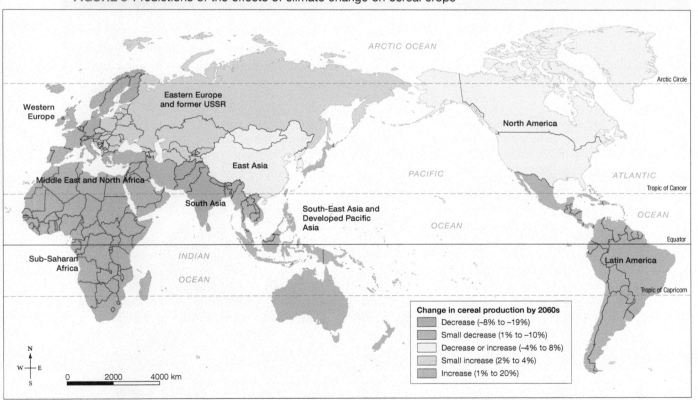

Source: Spatial Vision

1.2.5 Activities

To answer questions online and to receive **immediate feedback** and **sample responses** for every question, go to your learnON title at www.jacplus.com.au. *Note*: Question numbers may vary slightly.

Refer to figure 5.

1. Which regions of the world are predicted to experience decreases in cereal production as a result of climate *change*?
2. Which regions are predicted to show an increase in cereal production?
3. Some countries, particularly those with a shortage of arable land, are buying land in other countries specifically for food production. What do you see as the advantages and disadvantages of such land purchases?
4. Explain how each of the factors listed below may influence whether a country experiences a *change* in cereal production in the future as a result of climate *change*.
   (a) Level of development
   (b) Level of technological advancement
5. How would predictions help governments plan for future food security?

my **World** Atlas   **Deepen your understanding of this topic with related case studies and questions.**
   ⊙ **Change**

1.2.6 What is environment?

People live in and depend on the environment, so it has an important influence on our lives.

The biological and physical world that makes up the environment is important to us as a source of food and raw materials, a means of absorbing and recycling wastes, and a source of enjoyment and inspiration.

People perceive, adapt and use environments in many ways. For example, different people could look at a well-vegetated hillside and one might see it as a source of timber for construction, another might see a slope that could be cleared and terraced to produce food, while another might view it as a scenic environment for ecotourism.

FIGURE 6 The East Kolkata wetlands act as a sewage filtration system and recycle nutrients through the soil to allow a wide range of food crops to be grown. The ponds provide one-third of the city's fish supply and are a protected Ramsar site for migratory birds.

1.2.6 Activities

To answer questions online and to receive **immediate feedback** and **sample responses** for every question, go to your learnON title at www.jacplus.com.au. *Note*: Question numbers may vary slightly.

Refer to figure 6.

1. Look closely at the image and decide whether this is a natural or human *environment*. Justify your decision.
2. In what ways might people have *changed* or modified the natural *environment* to allow farming to take place?
3. How would the people and economy of Kolkata benefit from these wetlands?
4. What would you consider to be the advantages and disadvantages of using a wetland *environment* as a natural sewage treatment system?
5. A major threat to the future of the wetlands is the growing population and urban spread into this farming area. How might this affect Kolkata's:
   (a) food security
   (b) ability to deal with sewage and waste
   (c) internationally recognised wetland *environment*?

Deepen your understanding of this topic with related case studies and questions.
❯ **Environment**

1.2.7 What is sustainability?

Sustainability is about maintaining the capacity of the environment to support our lives and those of other living creatures.

Sustainability involves maintaining and managing our resources and environments for future generations. It is important to understand the causes of unsustainable situations to be able to make informed decisions on the best way to manage our natural world.

FIGURE 7 The unsustainable nature of fishing

1.2.7 Activities

To answer questions online and to receive **immediate feedback** and **sample responses** for every question, go to your learnON title at www.jacplus.com.au. *Note*: Question numbers may vary slightly.

Refer to figure 7.
1. What does this cartoon tell us about the *sustainable* nature of fishing?
2. How has modern technology contributed to a decline in world fish populations?
3. Suggests reasons why fishing can be considered an *unsustainable* practice.
4. Give reasons why some fish species are over-exploited while others are not.
5. What difficulties would arise with managing an ocean-based resource compared to a land-based resource?

myWorldAtlas
Deepen your understanding of this topic with related case studies and questions.
◉ Sustainability

1.2.8 What is space?

Everything has a location on the space that is the surface of the Earth. Studying the effects of location, the distribution of things across this space, and how the space is organised and managed by people, helps us to understand why the world is like it is.

A place can be described by its absolute location; for example: latitude and longitude, a grid reference, street directory reference or an address. Or, a place can be described using a relative location — where it is in relation to another place in terms of distance and direction.

FIGURE 8 Topographic map extract, Griffith, New South Wales

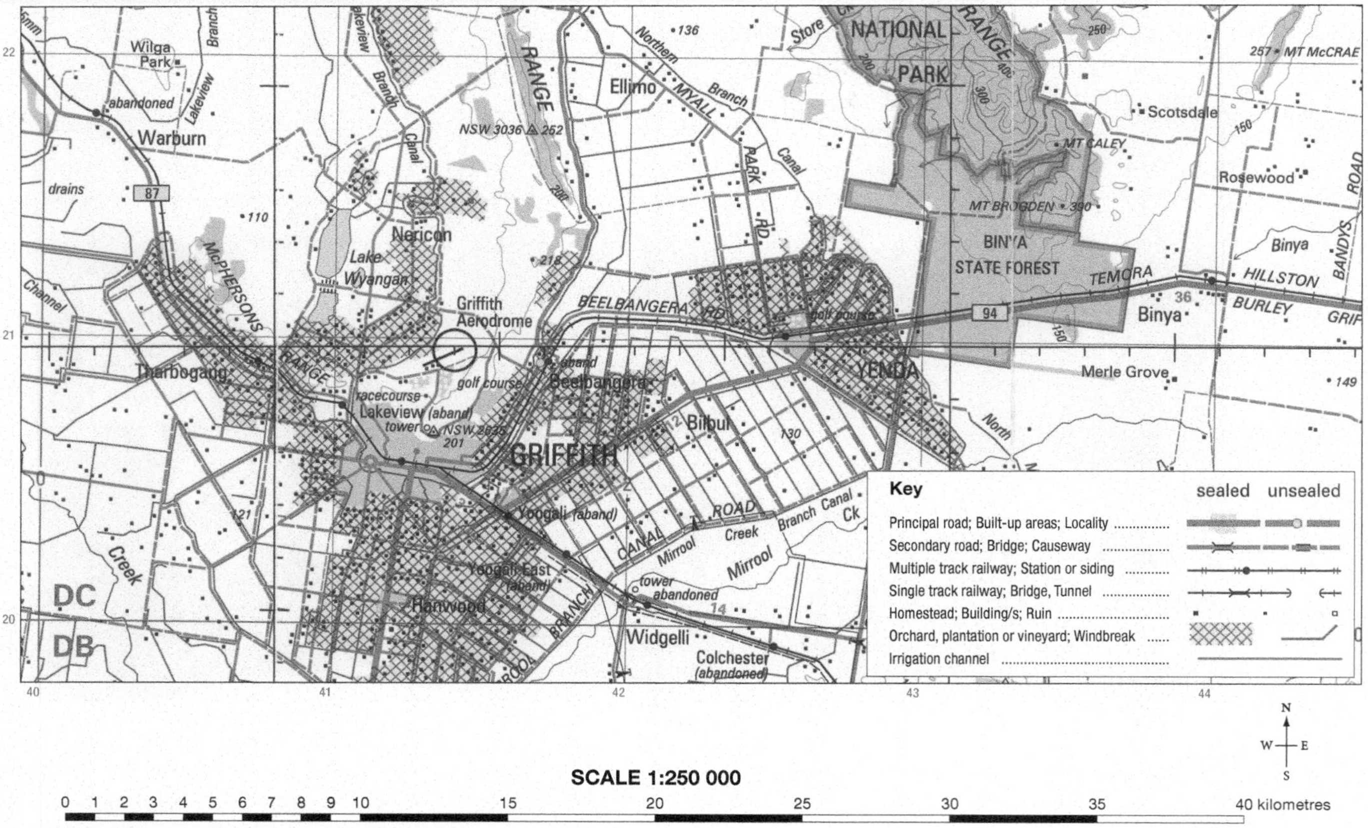

SCALE 1:250 000

0 1 2 3 4 5 6 7 8 9 10   15   20   25   30   35   40 kilometres

Source: Commonwealth of Australia (Geoscience Australia).

1.2.8 Activities

To answer questions online and to receive **immediate feedback** and **sample responses** for every question, go to your learnON title at www.jacplus.com.au. *Note*: Question numbers may vary slightly.

Refer to figure 8.

1. Use your atlas to locate Griffith. Provide its absolute location using longitude and latitude.
2. What is the location of Griffith Aerodrome relative to Lake Wyangan? Use distance and direction in your answer.
3. Identify the features located at the following grid references.
   (a) 413208
   (b) 435215
4. Give grid reference locations for the following.
   (a) Griffith Aerodrome
   (b) Lake Wyangan
5. Complete the following sentences by inserting the correct term to describe the distribution pattern. Choose from the following list of terms: rectangular; linear; scattered; clustered.
   (a) Buildings are distributed in a .... pattern.
   (b) Irrigated fields in the orchards and vineyards create a .... distribution pattern.
   (c) Railway lines form a .... pattern.
   (d) The distribution pattern for buildings around Yenda tends to be more .... than those to the west of Griffith.
6. How does the distribution of irrigated vineyards and orchards influence the distribution of buildings? Suggest a reason for your observation.
7. Describe how people have generally used the *space* shown in the map.

myWorldAtlas **Deepen your understanding of this topic with related case studies and questions.**
❂ **Space**

1.3 Review

1.3.1 Applying the concepts

Saudi Arabia is home to extensive desert regions. Today, thanks to advances in technology, much of the desert is being transformed into productive farming areas. Fruits, vegetables and grains are the main crops grown, and these help to improve the country's food security. Extensive drilling is tapping into underground aquifers as much as 1000 metres deep to access water for irrigation of water-hungry crops. Large circular sprays, called centre pivots, create a distinctive circular pattern of fields (see figure 1b).

Rainfall in the Wadi As-Sirhan Basin averages only 100–200 millimetres per year, which is insufficient to recharge underground aquifers. The water that is being pumped to the surface is actually 'fossil' water, possibly up to 20 000 years old. The volume of water that is being used for desert agriculture has more than tripled in just over 25 years.

FIGURE 1 Satellite images of the Wadi As-Sirhan Basin in Saudi Arabia. *Note:* Landsat imagery shows new vegetation as bright green, while dry vegetation or land lying fallow is shown as rust-coloured. Dry desert areas are shown as pink and yellow.

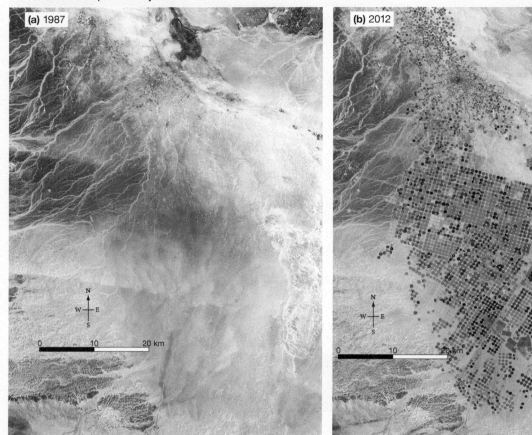

1.3 Activities

To answer questions online and to receive **immediate feedback** and **sample responses** for every question, go to your learnON title at www.jacplus.com.au. *Note*: Question numbers may vary slightly.

1. Where is Saudi Arabia located? (*space*)
2. Looking at figure 1a, how would you describe this *place*?
3. What do you think the white lines to the north-west of the image are? What does this tell you about the climate in this region? (*environment*, *space*)
4. Comparing the two images, describe the *changes* that irrigation has brought to this *environment*.
5. Each of the fields in figure 1b is approximately 1 kilometre wide. What does this indicate about the *scale* of this irrigation region?
6. How would the isolation of this irrigation region affect the movement of fresh produce to markets in cities? (*interconnection*)
7. Hydrologists (water engineers) believe that it will be economical to continue pumping water for only another 50 years. Is the use of groundwater *sustainable* in the future?

UNIT 1
BIOMES AND FOOD SECURITY

Food is essential to human life. To ensure we have reliable food sources, we alter our world biomes — clearing vegetation, diverting and storing water, adding chemicals and even changing landforms. We will need to carefully manage our limited land and water resources and use more sustainable farming practices to ensure we have future food security.

2 All the world is a biome   15

3 How can we feed the world?   36

4 What are the impacts of feeding our world?   67

5 Are we devouring our future?   88

6 2050 — food shortage or surplus?   107

7 Geographical inquiry: Biomes and food security   130

A traditional Asian food market in Malaysia

TOPIC 2
All the world is a biome

2.1 Overview

Numerous **videos** and **interactivities** are embedded just where you need them, at the point of learning, in your learnON title at www.jacplus.com.au. They will help you to learn the content and concepts covered in this topic.

2.1.1 Introduction

Where do the foods we eat and the natural products we use daily come from? Biomes. Biomes are communities of plants and animals that extend over large areas. Some are dense forests; some are deserts; some are grasslands, like much of Australia; and so the variations continue. Within each biome, plants and animals have similar adaptations that allow them to survive.

Biomes can be terrestrial (land based) or aquatic (water based). Understanding the diversity within them is essential to our survival and wellbeing.

Within each biome, there are many variations in the landscape and climate, and in the plants and animals that have adapted to survive there.

Starter questions

1. As a class, develop your own definition of the word *biome*.
2. What information about biomes is conveyed by the images shown on the previous page?
3. Brainstorm a list of biomes.
4. Select one of the biomes you listed in question 3, and create a mind map of the unique features you think it has.
5. What type of biome do you live in? If you cannot decide, why do you think that is?
6. Make a list of the things you use that have come from a biome. Compare your list with other students in your class.

INQUIRY SEQUENCE

2.1	Overview		15
2.2	**SkillBuilder:** Describing spatial relationships in thematic maps	online only	16
2.3	What is a biome?		17
2.4	How do we use the grassland biome?		20
2.5	What are coastal wetlands?		22
2.6	**SkillBuilder:** Constructing and describing a transect on a topographic map	online only	25
2.7	Why are coral reefs unique?		25
2.8	What are Australia's major biomes?		28
2.9	Why are biomes different?		30
2.10	How do we protect biomes?	online only	35
2.11	**Review**	online only	35

2.2 SkillBuilder: Describing spatial relationships in thematic maps

WHAT ARE SPATIAL RELATIONSHIPS IN THEMATIC MAPS?

A spatial relationship is the interconnection between two or more pieces of information in a thematic map, and the degree to which they influence each other's distribution in space. Describing these relationships helps us understand how one thing affects another.

FIGURE 1 Thematic map of Asia showing biomes

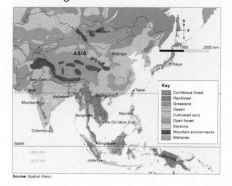

Source: Spatial Vision

Go online to access:

- a clear step-by-step explanation to help you master the skill
- a model of what you are aiming for
- a checklist of key aspects of the skill
- a series of questions to help you apply the skill and to check your understanding.

learn on RESOURCES — ONLINE ONLY

⊞ **Watch this eLesson:** Describing spatial relationships in thematic maps (eles-1726)

✦ **Try out this interactivity:** Describing spatial relationships in thematic maps (int-3344)

2.3 What is a biome?

2.3.1 What and where are the major biomes?

Biomes are sometimes referred to as ecosystems. They are places that share a similar climate and life forms. There are five distinct biomes across the Earth: forest, desert, grassland, tundra and aquatic biomes. Within each, there are variations in the visible landscape, and in the plants and animals that have adapted to survive in a particular climate.

FIGURE 1 Major biomes of the world

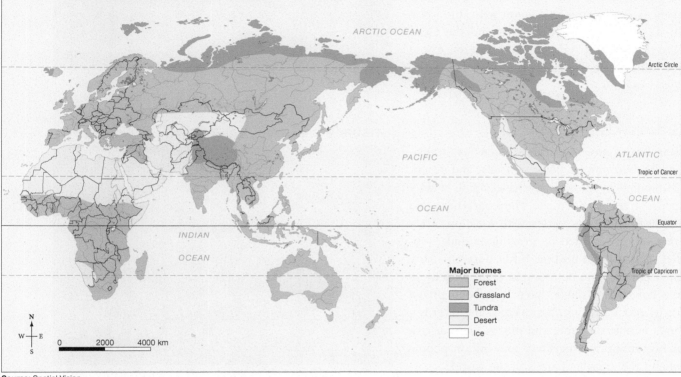

Major biomes
- Forest
- Grassland
- Tundra
- Desert
- Ice

Source: Spatial Vision

Forests

Forests are the most diverse ecosystems on the Earth. Ranging from hot, wet, tropical rainforests to temperate forests, they have an abundance of both plant and animal life. Over 50 per cent of all known plant and animal species are found in tropical rainforests. Forests are the source of over 7000 modern medicines, and many fruits and nuts originated in this biome. Forests help regulate global climate, because they absorb and use energy from the sun rather than reflect it back into the atmosphere. Forest plants recycle water back into the atmosphere, produce the oxygen we breathe, and store the carbon we produce. Forests are under threat from **deforestation**.

FIGURE 2 Forest

Deserts

Deserts are places of low rainfall and comprise the arid and semi-arid regions of the world. Generally they are places of temperature extremes — hot by day and cold by night. Most animals that inhabit deserts are nocturnal (active at night), and desert vegetation is sparse. Desert rain often evaporates before it hits the ground, or else it falls in short, heavy bursts. Following periods of heavy rain, deserts teem with life. Not all deserts are hot. Antarctica and the Gobi Desert in central Asia are cold deserts. About 300 million people around the world live in desert regions.

FIGURE 3 Desert

Grasslands

Grasslands are sometimes seen as transitional environments between forest and desert. Dominated by grass, they have small, widely spaced trees or no trees. The coarseness and height of the grass varies with location. They are mainly inhabited by grazing animals, reptiles and ground-nesting birds, though many other animals can be found in areas with more tree cover. Grasslands have long been prized for livestock grazing, but overgrazing of grasslands is unsustainable and places them at risk of becoming deserts. Over one billion people inhabit the grassland areas of the world.

FIGURE 4 Grassland

Tundra

Tundra is found in the coldest regions of the world, and lies beyond the **treeline**. The landscape is characterised by grasses, dwarf shrubs, mosses and lichens. The growing season is short. Tundra falls into three distinct categories — Arctic, Antarctic and alpine — but they share the common characteristic of low temperatures. In Arctic regions there is a layer beneath the surface known as permafrost — permanently frozen ground. The tundra biome is the most vulnerable to global warming, because plants and animals have little tolerance for environmental changes that reduce snow cover.

FIGURE 5 Tundra

Aquatic biomes

Water covers about three-quarters of the Earth and can be classified as fresh water or marine. Freshwater biomes contain very little salt and are found on land; these include lakes, rivers and wetlands. Marine biomes are the saltwater regions of the Earth and include oceans, coral reefs and estuaries. Marine environments are teeming with plant and animal life, and are a major food source. Elements taken from the roots of mangroves have been used in the development of cancer remedies. Compounds from other marine life have also been used in cosmetics and toothpaste.

FIGURE 6 Water

2.3 Activities

To answer questions online and to receive **immediate feedback** and **sample responses** for every question, go to your learnON title at www.jacplus.com.au. *Note*: Question numbers may vary slightly.

Remember
1. Name the five major biomes of the Earth and classify them as either aquatic or terrestrial.
2. Identify the broad characteristics biomes share.

Explain
3. Look carefully at figure 1. Using geographic terminology and concepts (including reference to latitude), describe the location and characteristics of the five major biomes.
4. (a) The map in figure 1 identifies five basic biomes. However, the text tells you that within each there are variations. Suggest reasons for these variations.
   (b) Are all biomes equally important? Explain your answer.

Predict
5. Select one of the categories of biomes described in this section. Suggest how this biome might be *changed* and used by humans and what impact this *change* might have on the *environment*.

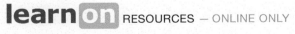

2.4 How do we use the grassland biome?

2.4.1 What are the characteristics of grasslands?

Grassland, pampas, savanna, chaparral, cerrado, **prairie**, rangeland and steppe all refer to a landscape that is dominated by grass. Once, grasslands occupied about 42 per cent of the Earth's land surface, but today they make up about 25 per cent of its land area. Grasslands are found on every continent except Antarctica.

The grassland biome is dominated by grasses, and generally has few or no trees, though there may be more tree cover in adjoining areas, such as along riverbanks. They develop in places where there is not enough rain to support a forest but too much rain for a desert; for this reason they are sometimes referred to as a transitional landscape.

Grasslands are found in both temperate and tropical areas where rainfall is between 250 mm and 900 mm per year. In tropical regions, grasslands tend to have a distinct wet and dry season. In temperate regions, the summers tend to be hot and the winters cool. Generally, grasslands in the southern hemisphere receive more rainfall.

Grasslands can occur naturally or as a result of human activity. The presence of large numbers of grazing animals and frequent fires prevent the growth of tree seedlings and promote the spread of grasses. Unlike other plant species, grasses can continue to grow even when they are continually grazed by animals, because their growth points are low and close to the soil. Because grasses are fast-growing plants, they can support a high density of grazing animals, and they regenerate quickly after fire.

Some grasses can be up to two metres in height, with roots extending up to one metre below the soil.

FIGURE 1 Grasslands occupy about a quarter of the Earth's land surface.

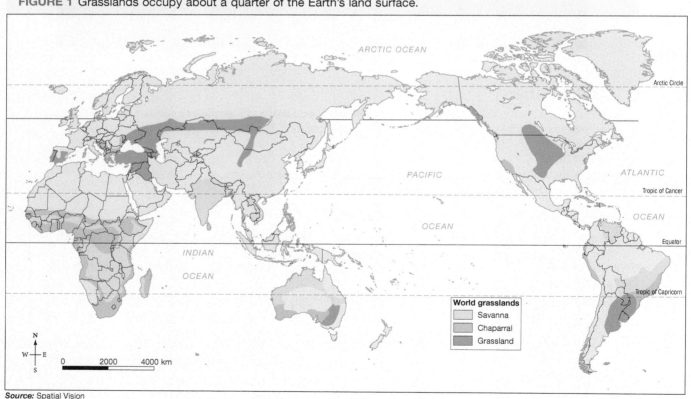

Source: Spatial Vision

2.4.2 Why are grasslands important?

Grasslands are the most useful biome for agriculture because the soils are generally deep and fertile. They are ideally suited for growing crops or creating pasture for grazing animals. The prairies of North America, for example, are one of the richest agricultural regions on Earth.

Grasslands are also one of the most endangered biomes and are easily turned into desert. The entire ecosystem depends on its grasses and their annual regeneration. It is almost impossible to re-establish a grassland ecosystem once desert has taken over.

Almost one billion people depend on grasslands for their livelihood or as a food source.

Grasslands often depend on fire to germinate their seeds and generate new plant growth. Indigenous populations, such as Australian Aboriginal peoples, used this technique to flush out any wildlife that was hidden by long grass.

In more recent times, grasslands have been used for livestock grazing and are increasingly under pressure from **urbanisation**. Grasslands have also become popular tourist destinations, because people flock to them to see majestic herds such as wildebeest, caribou and zebra, as well as the migratory birds that periodically inhabit these environments.

All the major food grains — corn, wheat, oats, barley, millet, rye and sorghum — have their origins in the grassland biome. Wild varieties of these grains are used to help keep cultivated strains disease free. Many native grass species have been used to treat diseases including HIV and cancer. Others have proven to have properties for treating headaches and toothache.

Grasslands are also the source of a variety of plants whose fibres can be woven into clothing. The best known and most widely used fibre is cotton. Harvested from the cottonseed, it is used to produce yarn that is then knitted or sewn to make clothing. Lesser known fibres include flax and hemp. Harvested from the stalk of the plant, both fibres are much sturdier and more rigid than cotton but can be woven to produce fabric. Hemp in particular is highly absorbent and has UV blocking qualities.

FIGURE 2 Grasslands can support a high density of grazing animals. In Australia, we use grasslands for fine wool production.

FIGURE 3 Wheat is a type of grass.

In Australia today, less than one per cent of native grasslands survive, and they are now considered one of the most threatened Australian habitats. Since European occupation, most native grassland has been removed or changed by farming and other development. Vast areas of grassland were cleared for crops, and introduced grasses were planted for grazing animals such as sheep and cattle.

2.4 Activities

To answer questions online and to receive **immediate feedback** and **sample responses** for every question, go to your learnON title at www.jacplus.com.au. *Note:* Question numbers may vary slightly.

Remember

1. What is a grassland?
2. Describe the global distribution of grasslands.

3. Why are grasslands an important *environment*?
4. Describe the major threats to this *environment*.

Explain

5. Explain how Indigenous populations used the grassland *environment* in a *sustainable* way.
6. Explain why so little of Australia's grasslands remain.

Discover

7. Grasslands are located on six of the Earth's seven continents. Working in groups, investigate one of the grassland biomes. Using ICT, create a presentation on your chosen biome that covers the following:
   (a) the characteristics of the *environment*, including climate and types of grasses that dominate this *place*
   (b) the animals that are commonly found there
   (c) how the *environment* is used and *changed* for the production of food, fibre and wood products
   (d) threats to this particular grassland, including the *scale* of these threats
   (e) what is being done to manage this grassland *environment* in a *sustainable* manner.
   Investigate how different factors involved in ethical decision-making can be managed by people and groups.

 Deepen your understanding of this topic with related case studies and questions.
◉ **Wheat**

 RESOURCES — ONLINE ONLY

 Try out this interactivity: Grass, grains and grazing (int-3318)

2.5 What are coastal wetlands?

2.5.1 What are wetlands?

Wetlands are biomes where the ground is saturated, either permanently or seasonally. They are found on every continent except Antarctica. Wetlands include areas that are commonly referred to as marshes, swamps and bogs. In coastal areas they are often tidal and are flooded for part of the day. In the past they were often considered a 'waste of space', and in developed nations they were sometimes drained for agriculture or the spread of urban settlements.

2.5.2 Are wetlands important?

Wetlands are a highly productive biome. They provide important habitats and breeding grounds for a variety of marine and freshwater species. In fact, a wide variety of aquatic species that we eat, such as fish, begin their life cycle in the sheltered waters of wetlands. They are also important nesting places for a large number of migratory birds.

Wetlands are also a natural filtering system and help purify water and filter out pollutants before they reach the coast. In addition, they help regulate river flow and stabilise the shoreline. Figure 1 shows a cross-section through a mangrove wetland.

FIGURE 1 Cross-section of a wetland

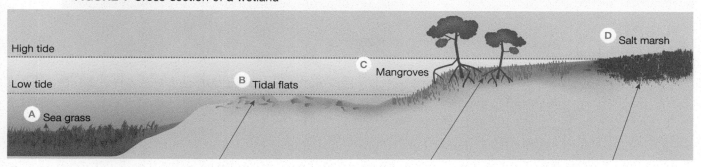

High tide

Low tide

A Sea grass

B Tidal flats

C Mangroves

D Salt marsh

A Seagrass meadows:
- are covered by water all the time
- bind the mud and provide shelter for young fish
- produce **organic matter**, which is consumed by marine creatures.

B Tidal flats:
- are covered by tides most of the time
- are exposed for short periods of the day (low tide)
- are formed by silt and sand that has been deposited by tides and rivers
- provide a feeding area for birds and fish.

C Mangroves:
- have **pneumatophores** that trap sediment and pollutants from the land and sea
- change shallow water into swampland
- store water and release it slowly into the ecosystem
- have leaves that decompose and provide a food source for marine life
- provide shelter, breeding grounds and a nursery for marine creatures and birds.

D Salt marshes:
- are covered by water several times per year
- provide decomposing plant matter — an additional food source for marine life
- have high concentrations of salt.

FIGURE 2 Seagrass

FIGURE 3 Pneumatophores

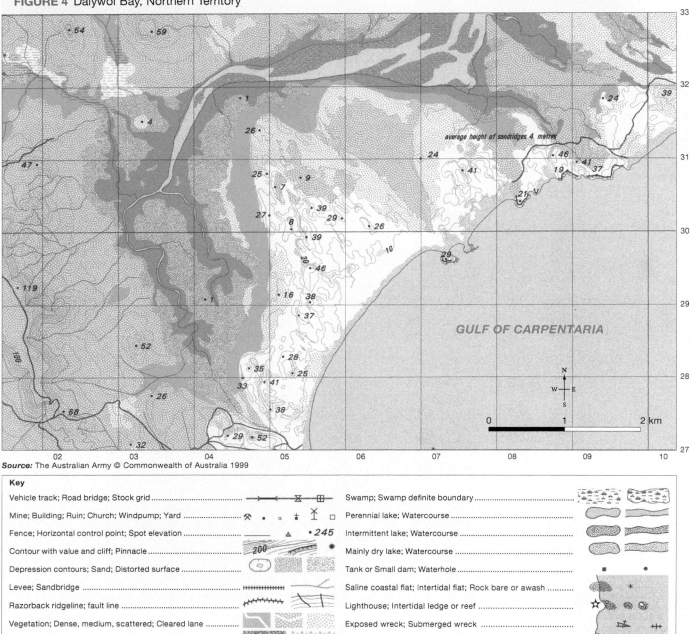

FIGURE 4 Dalywoi Bay, Northern Territory

Source: The Australian Army © Commonwealth of Australia 1999

Key

Vehicle track; Road bridge; Stock grid	Swamp; Swamp definite boundary
Mine; Building; Ruin; Church; Windpump; Yard	Perennial lake; Watercourse
Fence; Horizontal control point; Spot elevation	Intermittent lake; Watercourse
Contour with value and cliff; Pinnacle	Mainly dry lake; Watercourse
Depression contours; Sand; Distorted surface	Tank or Small dam; Waterhole
Levee; Sandbridge	Saline coastal flat; Intertidal flat; Rock bare or awash
Razorback ridgeline; fault line	Lighthouse; Intertidal ledge or reef
Vegetation; Dense, medium, scattered; Cleared lane	Exposed wreck; Submerged wreck
Rain forest; Pine	Submerged reef; Submerged rock
Orchard or vineyard; Line if trees or windbreak	Indefinite watercourse; Mangrove swamp
Watercourse; Area subject to inundation	

2.5 Activities

To answer questions online and to receive **immediate feedback** and **sample responses** for every question, go to your learnON title at www.jacplus.com.au. *Note:* Question numbers may vary slightly.

Explain

1. How have mangroves adapted to survive in their marine–terrestrial *environment*?
2. What are seagrass meadows and why are they important?
3. Refer to figure 4 and describe the landscape at the following grid references: 042309, 071329, 030320 and 055290.

2.6 SkillBuilder: Constructing and describing a transect on a topographic map

WHAT IS A TRANSECT?

A transect is a cross-section with additional detail, which summarises information about the environment.

Go online to access:

- a clear step-by-step explanation to help you master the skill
- a model of what you are aiming for
- a checklist of key aspects of the skill
- a series of questions to help you apply the skill and to check your understanding.

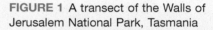

FIGURE 1 A transect of the Walls of Jerusalem National Park, Tasmania

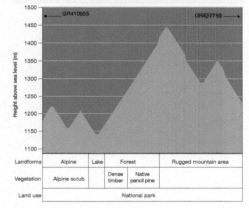

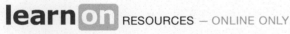

learn on RESOURCES — ONLINE ONLY

Watch this eLesson: Constructing and describing a transect on a topographic map (eles-1727)

Try out this interactivity: Constructing and describing a transect on a topographic map (int-3345)

2.7 Why are coral reefs unique?

2.7.1 Formation of coral reefs

Coral reefs are found in spaces around tropical and subtropical shores. They require specific temperatures and sea conditions and an area free from sediment. Coral reefs are built by tiny animals called **coral polyps**. The upper layer is alive, growing on the remains of millions of dead coral. Coral reefs are one of the oldest ecosystems on Earth and also one of the most vulnerable to human activity.

Coral reefs are one of the most diverse biomes on Earth and are built by polyps that live in groups. A reef is a layer of living coral growing on the remains of millions of layers of dead coral. There are inner and outer reefs as well as coral cays (small islands of coral) and coral atolls.

FIGURE 1 Anatomy of a coral reef

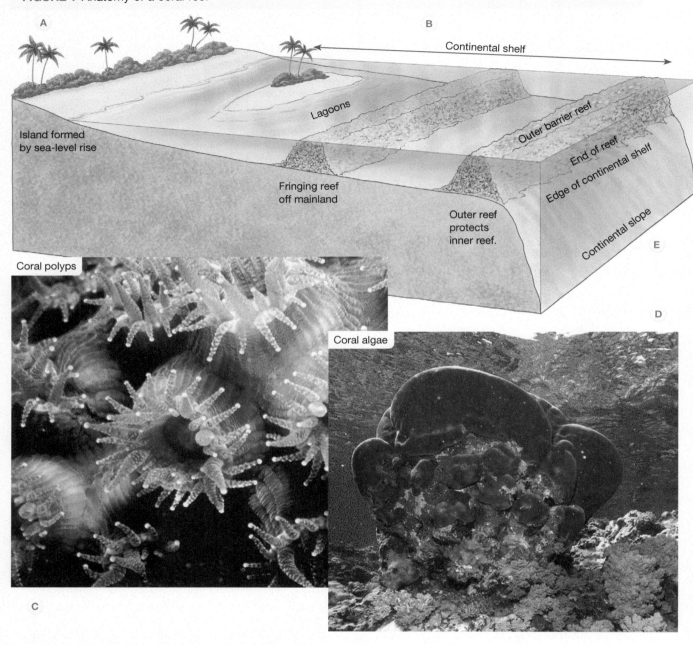

A Continental island and fringing reef

B • Corals form in warm shallow salt water where the temperature is between 18 °C and 26 °C.

   • Water must be clear, with abundant sunlight and gentle wave action to provide oxygen and distribute nutrients.

C Coral polyps have soft, hollow bodies shaped like a sac with tentacles around the opening. They cover themselves in a limestone skeleton and divide and form new polyps.

D Producers such as algae give coral its colour and provide a food source for marine life such as fish. Coral reefs support at least one-third of all marine species. They are the marine equivalent of the tropical rainforest.

E Beyond the continental shelf, the water is too deep and cold for coral. Sunlight cannot penetrate to allow coral growth.

2.7.2 Benefits of coral reefs

Today, about 500 million people rely on reef systems, either for their livelihood, as a source of food, or as a means of protecting their homes along the coastline. (Coral reefs help break up wave action, so waves have less energy when they reach the shoreline, thus reducing coastal erosion.) It is estimated that coral reefs contribute between US$28.8 billion and US$375 billion to the global economy each year.

Reefs are important to both the fishing and tourism industries. Approximately 2 million people visit Australia's Great Barrier Reef Marine Park alone, generating more than A$2 billion for the local economy. Nearly one-third of all tourists who visit Australia also visit the Great Barrier Reef.

Coral reefs have been found to contain compounds vital to the development of new medicines.
- Painkillers have been developed from the venom of cone shells.
- Some cancer treatments come from algae.
- Treatments for cardiovascular disease and HIV include compounds that were originally found in coral reefs.

2.7.3 Threats to coral reefs

Reefs also face a variety of threats.
- Urban development requires land clearing and wetland drainage, which increases erosion. Sediment washed into water prevents sunlight penetrating the water.
- Contamination by fossil fuels, chemical waste and agricultural fertilisers pollutes the sea.
- Tourism damages coral through boats dropping anchor or tourists directly removing coral or walking on it.
- Global warming increases water temperature, which bleaches the coral, turning it white and destroying the reef system.
- Predators, such as the crown of thorns starfish prey on coral polyps, which affects the whole ecosystem.

FIGURE 2 The formation of fringing reefs, barrier reefs and coral atolls

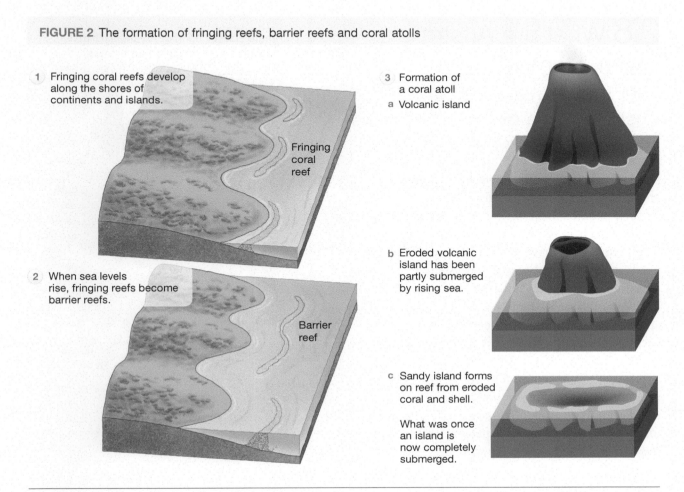

1 Fringing coral reefs develop along the shores of continents and islands.

Fringing coral reef

2 When sea levels rise, fringing reefs become barrier reefs.

Barrier reef

3 Formation of a coral atoll

a Volcanic island

b Eroded volcanic island has been partly submerged by rising sea.

c Sandy island forms on reef from eroded coral and shell.

What was once an island is now completely submerged.

2.7 Activities

To answer questions online and to receive **immediate feedback** and **sample responses** for every question, go to your learnON title at www.jacplus.com.au. *Note*: Question numbers may vary slightly.

Explain

1. Explain the difference between a fringing reef and a barrier reef.
2. With the aid of a diagram, explain how coral reefs are formed.
3. Why are coral reefs important *places*?

Discover

4. Investigate two of the threats to coral reefs and prepare an annotated visual display that outlines:
   - the nature of the threat
   - the *changes* that will occur or have occurred as a result of this threat
   - the impact of these *changes* on the *environment*, including references to the rate and *scale* of the *changes*
   - a strategy for the long-term *sustainable* management of the reef *environment*.

Predict

5. Coral reefs are highly susceptible to climate *change*.
   (a) Explain what you understand by the term *climate change*.
   (b) With the aid of a diagram, explain how you think the coral reef *environment* would *change* if sea temperatures were to rise by 2 °C.

---

myWorldAtlas   Deepen your understanding of this topic with related case studies and questions.
◊ The Great Barrier Reef

2.8 What are Australia's major biomes?

2.8.1 What factors shape Australian biomes?

Australia is a land of contrasts. Its mountain ranges and river systems are small by world standards. In the north are tropical rainforests and savanna grasslands. In the centre is a wide expanse of desert that is second in area only to the Sahara Desert in Africa. In the south, temperate forests and grasslands dominate. Australia also has vast wetlands and coastal ecosystems.

Before European colonisation, the Australian landscape was shaped largely by natural processes and Aboriginal burning practices. With European occupation came large-scale land clearing, irrigation of the land through water diversion from rivers, and the draining of wetlands. New plant and animal species were introduced to the landscape. However, despite the large-scale changes made since European occupation, Australia's major biomes are still clearly evident.

FIGURE 1 Climate classification of Australia. Climate plays a major role in the development of biomes.

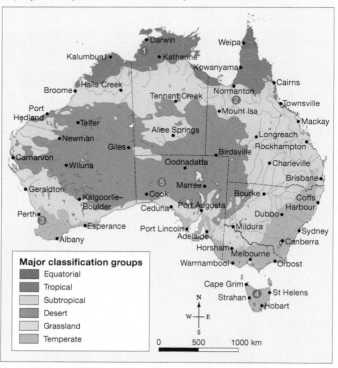

Major classification groups
- Equatorial
- Tropical
- Subtropical
- Desert
- Grassland
- Temperate

Source: Data copyright Commonwealth of Australia, 2013 Bureau of Meteorology

---

① Wetlands and rivers

In northern Australia, wetlands have been inhabited by Indigenous Australian peoples since the beginning of the Dreaming (more than 50 000 years). These areas provided them with food and water, and they used wetland plants such as river reeds and lily leaves to make fishing traps. Today, wetlands are still important habitats for native and migratory birds. In many parts of Australia, they are under threat, because water is diverted from rivers to produce food crops and cotton.

② Savanna (grasslands)

Grasslands (or savanna) are generally flat, having either few trees and shrubs or very open woodland. For many native species, grasslands provide vital habitat and protection from predators. Many grasslands depend on a regular cycle of burning to germinate their seeds and to revive the land. Periodic burning also prevents trees from gaining dominance in the landscape. Before European occupation, Indigenous Australian peoples hunted the animals in the grasslands. However, since then, grasslands have been used extensively for grazing. These areas often mark the transition between desert and forest, and are a very fragile biome. Without careful management they can quickly change to desert. Less than one per cent of Australia's original native grasslands survive today.

③ Seagrass meadows

Seagrasses are submerged flowering plants that form colonies off long, sandy ocean beaches, creating dense areas that resemble meadows. Of the 60 known species of seagrass, at least half are found in Australia's tropical and temperate waters. Western Australia alone is home to the largest seagrass meadow in the world. Seagrasses provide important habitats for a wide variety of marine creatures, including rock lobsters, dugongs and sea turtles. They also absorb nutrients from coastal run-off, slow water flow, help stabilise sediment, and keep water clear.

④ Old-growth forest

An old-growth forest is one in its oldest growth stage. It is multi-layered, and the trees are of mixed ages. Generally, there are few signs of human disturbance. These forests are biologically diverse, often home to rare or endangered species, and show signs of natural regeneration and decomposition. The trees within some old-growth forests have been felled for their timber and to create paper products. **Logging** can reduce **biodiversity**, affecting not only the forest itself but also the indigenous plant and animal species that rely on the old-growth habitat.

FIGURE 2 A billabong in Kakadu National Park, part of an extensive wetland system that develops during the wet season

FIGURE 3 Savanna biomes are typically dominated by grasses and scattered trees.

FIGURE 4 Seagrass meadows provide food, shelter and breeding grounds for marine life.

FIGURE 5 Different layers of vegetation can be seen in old-growth forests.

It is estimated that **clearfelling** of Tasmania's old-growth forests would release as much as 650 tonnes of carbon per hectare into the atmosphere. In Victoria, near Melbourne, many old-growth forests lie within protected water supply catchments and help maintain the integrity of the city's water supply.

⑤ Desert

Australian deserts are places of temperature extremes. During the day, temperatures sometimes exceed 50 °C, but at night this can drop to freezing. Australia's desert regions are often referred to as the outback but they are not all endless plains of sand. Some, such as the Simpson and Great Sandy Deserts, are dominated by sand. The Nullarbor Plain and Barkly Tablelands are mainly smooth and flat, while the Gibson Desert and Sturt Stony Desert contain low, rocky hills. In some areas, the landscape is dominated by spinifex and acacia shrubs.

FIGURE 6 Vegetation in desert biomes has specific adaptations to enable it to survive in the harsh climate.

2.8 Activities

To answer questions online and to receive **immediate feedback** and **sample responses** for every question, go to your learnON title at www.jacplus.com.au. *Note:* Question numbers may vary slightly.

Explain

1. Explain why Australia has such a wide variety of biomes.
2. With the aid of a flow diagram, show how the Australian *environment changed* when European occupiers arrived.
3. Explain what you understand by the term *biodiversity* and why it is important.
4. What other types of biomes would you expect to find in Australia?

Discover

5. Select one of the climate zones shown in figure 1 and investigate the biomes found within it. Prepare a report on the importance of one of these biomes and how it has *changed* over time. What do you think should be done to protect it?
6. (a) Investigate one of the Australian biomes and examine how plants and animals have adapted to survive in it. Create a class collage depicting the way plants and animals have adapted to the Australian *environment*.
   (b) Explore what this biome is like in another *place* on Earth. With the aid of a Venn diagram, compare the two.

Predict

7. (a) Describe the *interconnection* between biomes and climate.
   (b) Select one of the biomes discussed in this section. Predict what might happen if it were *changed*; for example, if the wetlands were drained, or all the old-growth forests were cut down. Include a reference to the effect this would have on biodiversity. Produce a before-and-after snapshot of your chosen biome.

Deepen your understanding of this topic with related case studies and questions.
⊙ **Australia's alpine biomes**

2.9 Why are biomes different?

2.9.1 Why is climate the major influence on biomes?

Biomes are controlled by climate. In turn, climate is influenced by factors such as distance from the equator; altitude and distance from the sea; the direction of prevailing winds; and the location of mountain ranges. These play a key role in determining a region's climate and soil, which ultimately influence which plants and animals will inhabit it.

Temperature and rainfall patterns across the Earth determine which plant and animal species can survive in a particular biome. For instance, a polar bear could not survive in the hot climate of a desert or a tropical rainforest. Camels would not survive in the polar regions of the Earth, and fish cannot survive without water. The plants and animals of a region have adapted over time to the variations in the region's climate conditions.

2.9.2 What are the major influences on climate?

The major influences on climate are the geographical features of the Earth's surface, such as mountain ranges and **latitude**. Similarities have been found in the adaptations of plant and animal species in mountain regions and those found near the poles.

Landform

The major geographical influence on climate is the location of mountain ranges (see figure 1). Mountain ranges affect the amount of **precipitation** that reaches inland areas, because they pose a barrier to moisture-laden prevailing winds. **Rain shadows** form on the **leeward** side of mountains (opposite to the **windward** side). Deserts often form in rain shadows.

Altitude also plays a significant role in determining climate. Temperatures fall by 0.65 °C for every 100 metres increase in elevation. This can be illustrated by Mt Kilimanjaro (figure 2), which is located on the border of Tanzania and Kenya, in Africa, approximately 3° latitude from the equator. Towering 5895 metres above sea level, Mt Kilimanjaro is the highest mountain in Africa. Depending on the time of the day, the temperature at the base of the mountain ranges from 21 °C to 27 °C. At the summit, temperatures can plummet to −26 °C. As you move from base to summit, variations occur in the landscape as it transitions from rainforest to alpine desert to desert tundra.

Latitude

The sun's rays are more direct at the equator. With more energy focused on that region, it heats up more quickly. At the poles, the sun's rays are spread over a larger area and therefore cannot heat up as effectively. As a result, areas at the poles are much cooler than areas at the equator (see figure 3).

The tilt of the Earth on its axis also has a part to play. When a hemisphere tilts towards

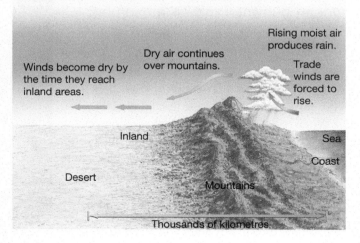

FIGURE 1 The influence of mountains on climate. This illustration shows the pattern typical on the east coast of Australia, where there are warm ocean currents.

FIGURE 2 Mt Kilimanjaro is only three degrees south of the equator but it is 5895 metres high; its altitude is the reason it has snow on its summit.

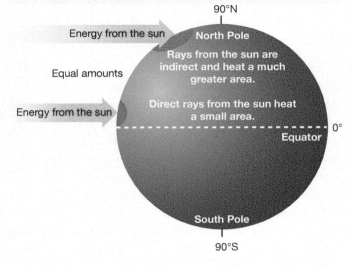

FIGURE 3 The influence of latitude on climate. The rotation of the Earth around the sun and the tilt of the Earth on its axis also influence the seasons.

the sun, the sun's rays hit it more directly. This means that a larger space is in more intense sunlight for longer. The days are longer and warmer, and the hemisphere experiences summer. The reverse is true when a hemisphere tilts away from the sun in winter.

Ocean currents and air movement

There are other factors that influence climate and play a role in the development of biomes. Two of these are ocean currents and air movement. In addition, differences also occur when you move from the coast to inland areas.

When cold ocean currents flow close to a warm land mass, a desert is more likely to form. This is because cold ocean currents cool the air above, causing less evaporation and making the air drier. As this air moves over the warm land, it heats up, making it less likely to release any moisture it holds; thus, deserts form. For example, cold ocean currents flow off the coast of Western Australia, while the Pacific Ocean currents are warmer. As a result, Perth on average receives less rainfall than Sydney.

2.9.3 The role of soil in biomes

Soil is important in determining which plants and animals inhabit a particular biome. Soils not only vary around the world but also within regions. The characteristics of soil are determined by:
• temperature
• rainfall
• the rocks and minerals that make up the bedrock, which is the basis of soil development.

The amount of vegetation present also plays an important role in determining the quality of the soil. Figure 4 shows a typical soil profile. The different soil layers are referred to as horizons.

Why do soils differ?

Biomes located in the high latitudes (those farthest from the equator) have lower temperatures and less exposure to sunlight than biomes located in the low latitudes (those close to the equator). There are also variations in the amount of precipitation that biomes receive. This is determined partly by their location in relation to the equator (see figure 5).

FIGURE 4 A typical soil profile has a number of distinct layers.

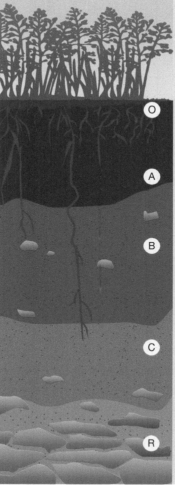

Horizon O (organic matter): A thin layer of decomposing matter, humus, and material that has not started to decompose, such as leaf litter

Horizon A (topsoil): The upper layer of soil, nearest the surface. It is rich in nutrients to support plant growth and usually dark in colour. Most plant roots and soil organisms are found in this horizon, which will also contain some minerals. In areas of high rainfall, such as tropical rainforests, minerals will be leached out of this layer. A constant supply of decomposing organic matter is needed to maintain soil fertility.

Horizon B (subsoil): Plant litter is not present in horizon B; as a result, little humus is present. Nutrients leached from horizon A accumulate in this layer, which will be lighter in colour and contain more minerals than the horizon above.

Horizon C (parent material): Weathered rock that has not broken down far enough to be soil. Nutrients leached from horizon A are also found in this layer. It will have a high mineral content; the type is determined by the underlying bedrock.

Horizon R (bedrock): Underlying layer of partly weathered rock

FIGURE 5 Just as there is a link between climate and latitude, there is also a link between soil, climate and latitude.

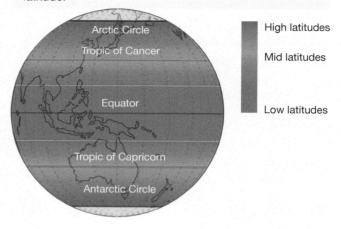

Temperature and precipitation patterns are important factors in determining the rate of soil development. However, soil moisture, its nutrient content and the length of the growing season also play key roles in soil development and, ultimately, the biodiversity of a biome.

Soil is more abundant in biomes that have both high temperatures and high moisture than in cold, dry regions. This is because erosion of bedrock is more rapid when moisture content is high, and organic material decomposes at a faster rate in high temperatures. The decomposition of organic matter provides the nutrients needed for plant growth, which in turn die and decompose in a continuous cycle. This is further demonstrated in figure 6.

FIGURE 6 Different biomes have different soil and vegetation characteristics.

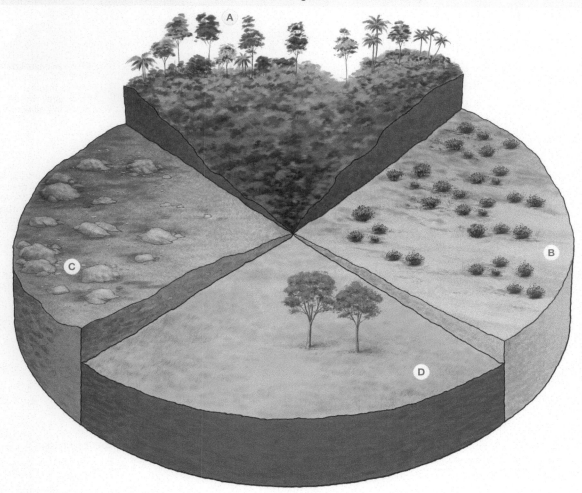

A **Tropical rainforest**
- High temperatures cause weathering, or breakdown, of rocks and organic matter.
- High rainfall **leaches** nutrients from the soil.
- Soil is often reddish because of high iron levels.
- Organic matter is often a shallow layer on the surface. Nutrients are constantly recycled, allowing the rainforest to flourish.
- Soil fertility is rapidly lost if trees are removed, as the supply of organic material is no longer present.

B **Desert**
- Limited vegetation means a limited supply of organic material for soil development.
- High temperatures rapidly break down any organic material.
- Soils are pale in colour rather than dark.
- Lack of rainfall limits plant growth.
- Lack of vegetation makes surface soil unstable and easily blown away.
- Soil does not have time to develop and mature.

C **Tundra**
- Soil is shallow and poorly developed.
- Includes layers that are frozen for long periods.
- Subsoil may be permanently frozen.
- Is covered by ice and snow for most of the year.
- Growing season may be limited to a few weeks.
- Soil may contain large amounts of organic material but extreme cold means it breaks down very slowly.
- Trees are absent; mosses and stunted grasses dominate.

D **Temperate**
- Generally brown in colour, soils have distinctive horizons and are generally about one metre deep.
- Ideal soils for agriculture; they are not subjected to the extremes of climate found in high and low latitudes.
- Moderate climate; temperature and rainfall are sufficient for plant growth.
- Dominated by temperate grasslands and deciduous forests.

What else is in the soil?

Soil not only supports the plants and animals that we see on the surface of the land; the soil itself is also home to a variety of life forms such as bacteria, fungi, earthworms and algae.

While most soil organisms are too small to be seen, there are others that are visible. For instance, more than 400 000 earthworms can be found on a hectare of land. Regardless of size, all soil organisms play a vital role in maintaining soil quality and fertility. For example, earthworms:

- compost waste and fertilise the soil
- improve drainage and aeration
- bring subsoil to the surface and mix it with topsoil
- secrete nitrogen and chemicals that help bind the soil.

FIGURE 7 There are more microbes in a teaspoon of soil than there are people on Earth.

2.9 Activities

To answer questions online and to receive **immediate feedback** and **sample responses** for every question, go to your learnON title at www.jacplus.com.au. *Note*: Question numbers may vary slightly.

Remember

1. What are the major influences on the development of biomes?

Explain

2. Explain the difference between the windward and leeward side of a mountain range.
3. Explain why soils vary across biomes.

Predict

4. (a) Use your atlas to locate Rwanda in central Africa. What type of biome would you expect to find in Rwanda? Give reasons for your answer.
   (b) What do you think the soil would be like in Rwanda?
   (c) Use the internet to test your theory.
5. Would you expect to find soil variations within biomes? Give reasons for your answer.
6. Predict the *changes* that might occur if earthworms or micro-organisms within the soil were no longer present.

Think

7. (a) What type of climate and biomes would you expect to find at the equator? Why?
   (b) Describe the climate and landscape on Mount Kilimanjaro.
   (c) What are some of the factors that create variations in biomes?

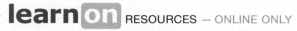

2.10 How do we protect biomes?

Access this subtopic at **www.jacplus.com.au**

2.11 Review

2.11.1 Review

The Review section contains a range of different questions and activities to help you revise and recall what you have learned, especially prior to a topic test.

2.11.2 Reflect

The Reflect section provides you with an opportunity to apply and extend your learning.

Access this subtopic at **www.jacplus.com.au**

TOPIC 3
How can we feed the world?

3.1 Overview

Numerous **videos** and **interactivities** are embedded just where you need them, at the point of learning, in your learnON title at www.jacplus.com.au. They will help you to learn the content and concepts covered in this topic.

3.1.1 Introduction

Food dominates every person's life. For many people, what to have for breakfast, lunch and dinner can be a constant thought and sometimes a worry.

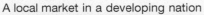

A local market in a developing nation

Starter questions

1. Have you ever thought about where the food you eat is actually produced? Do you know how much is produced in Australia and how much is imported? Check the labels of various foods in your pantry or fridge to work out the sources of your favourite or most consumed foods.
2. Do you think there is enough food in the world for everybody? Why? How do you know?
3. If you had to, could you grow fruit and vegetables in your home garden? Does your family have a garden and, if so, does your family grow its own food? Conduct a class discussion to establish how many students in your class grow their own food.

INQUIRY SEQUENCE

3.1 Overview		36
3.2 How can we feed the world?		37
3.3 What does the world eat?		40
3.4 How does traditional agriculture produce food?		43
3.5 How did Indigenous Australian peoples achieve food security?		46
3.6 SkillBuilder: Constructing ternary graphs	online only	50
3.7 How have we increased our food?		51
3.8 How are biomes modified for agriculture?		53
3.9 How is food produced in Australia?		54
3.10 What does a farming area look like?		59
3.11 SkillBuilder: Describing patterns and correlations on a topographic map	online only	61
3.12 Why is rice an important food crop?		62
3.13 Why is cacao a special food crop?	online only	66
3.14 Daly River: a sustainable ecosystem?	online only	66
3.15 How can aquaculture improve food security for Indigenous Australian peoples?	online only	66
3.16 Review	online only	66

3.2 How can we feed the world?

3.2.1 What are our food problems?

At the beginning of the twentieth century, the entire world population was less than 2 billion people. Today, the current world population is more than 7 billion. Earth's population is projected to rise to 9 billion people by 2050, and we all need food. What can we do to ensure there is enough food for everyone?

The map in figure 1 (on page 38) shows that crops occupy half the available agricultural land space. Almost all future population growth will occur in the developing world. This increased population, combined with higher standards of living in developing countries, will create enormous strains on land, water, energy and other natural resources.

There is currently about one-sixth of a hectare of **arable** land **per capita** in East and South Asia. With population growth, and almost no additional land available for agricultural expansion, arable land per capita will continue to decline.

FIGURE 1 World distribution of cropland, pasture and maize. More maize, for example, could be grown if improvements were made to seeds, irrigation, fertiliser and markets.

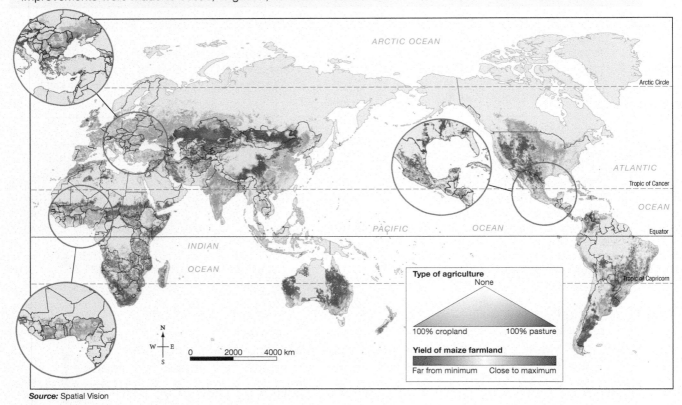

Source: Spatial Vision

3.2.2 Food production increases

Agricultural yields vary widely around the world owing to climate, management practices and the types of crops grown. Globally, 15 million square kilometres of land are used for growing crops — altogether, that's about the size of South America. Approximately 32 million square kilometres of land around the world are used for pasture — an area about the size of Africa. Across the Earth, most land that is suitable for agriculture is already used for that purpose and, in the past 50 years, we have increased our food production.

According to the Food and Agriculture Organization (FAO), the three main factors that have affected recent increases in world crop food production are:

- increased cropland and range-land area
- increased yield per unit area
- greater cropping intensity.

Current FAO projections suggest that cereal demand will increase by almost 50 per cent by 2050. This can either be obtained by increasing yields, expanding cropland through conversion of natural habitats, or growing crops more efficiently. Figure 2 shows the growth in crop yields from 1961 to what is proposed in 2030. Rice, maize and wheat have had significant increases in yield.

FIGURE 2 Crop yields in developing countries, 1961–2030

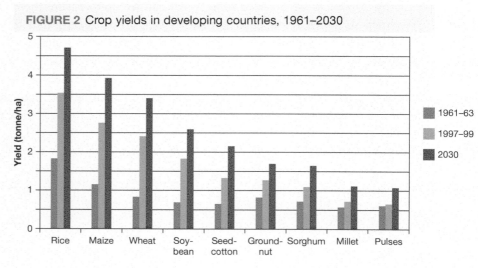

Agricultural **innovations** have also changed and increased global food production. They have boosted crop yields through advanced seed genetics; agronomic practices (scientific production of food plants); and product innovations that help farmers maximise productivity and quality. In this way, the nutritional content of crops can be increased (see figure 3).

3.2.3 We could do more

It should be possible to get more food out of the land we are already using. Figure 1 shows the places where maize yields could increase and become more **sustainable** by improving nutrient and water management, seed types and markets.

FIGURE 3 Farmers in a village in Kenya examine information on plant diseases using a laptop at a plant health clinic. They can also consult a plant pathologist and show them samples of their crops.

3.2 Activities

To answer questions online and to receive **immediate feedback** and **sample responses** for every question, go to your learnON title at www.jacplus.com.au. *Note:* Question numbers may vary slightly.

Remember

1. Refer to figure 1 and describe the distribution of *places* in the world with pasture and grasslands.
2. How could crop production be increased in *places* such as Eastern Europe or Western Africa?

Explain

3. Explain the impact of an increasing population on world *environments*.
4. Explain why agricultural innovations can *change* food production.

Discover

5. Research the reasons why the *environments* of Canada, Northern Africa and Central Australia, shown on figure 1, do not produce any crops.
6. Figure 1 shows where more crops could be grown. Investigate how Mexico or a country in West Africa or Eastern Europe could improve the *sustainability* of its agriculture.

Predict

7. With reference to specific *places*, suggest how increasing population densities might influence future crop production.
8. Figure 1 refers to the potential increase in maize crop yields. Suggest how this could be of benefit to a future world population.

Think

9. Should countries in the developed world be supporting those who struggle to produce their own food?
10. Would food production be secure if we grew fewer crops better?
11. Use the **Feed the world** weblink in Resources tab to watch the interactive maps. Describe how the challenge of meeting the needs of a growing and increasingly affluent population can be met.
12. Figure 2 refers to crop yields in developing countries over time. Suggest why rice, maize and wheat have the greatest increases in yields. Would these increases be similar in the developed regions of the world? Explain your answer.

 RESOURCES — ONLINE ONLY

🔗 **Explore more with this weblink:** Feed the world

3.3 What does the world eat?

3.3.1 The major food staples

Staple foods are those that are eaten regularly and in such quantities that they constitute a dominant portion of a diet. They form part of the normal, everyday meals of the people living in a particular place or country. They are called staples because they are easy to access and are grown or produced locally.

The world has over 50 000 **edible** plants. Staple foods vary from place to place, but are typically inexpensive or readily available. The staple food of an area is normally interconnected to the climate of that area and the type of land.

Most staple foods are cereals, such as wheat, barley, rye, oats, maize and rice; or root vegetables, such as potatoes, yams, taro and cassava. Rice, maize and wheat provide 60 per cent of the world's food energy intake; 4 billion people rely on them as their staple food.

Other staple foods include legumes, such as soya beans and sago; fruits, such as breadfruit and plantains (a type of banana); and fish.

FIGURE 1 Staple foods around the world

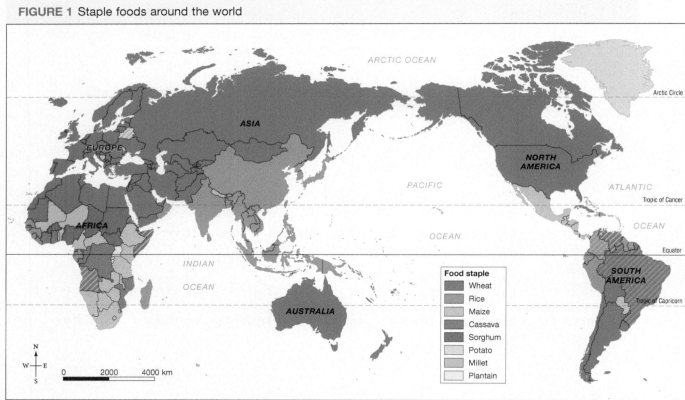

Source: Data from FAO

3.3.2 Wheat, maize and fish

Wheat is a cereal grain that is cultivated across the world. In 2015, world production of wheat was 715.90 million tonnes, making it the third most produced cereal after maize (1018.1 million tonnes) and rice (740.9 million tonnes). World trade in wheat is greater than for all other crops combined. In terms of total production tonnage, it is currently second to rice as the main human food crop and ahead of maize, after allowing for maize's more extensive use as an animal feed.

Wheat was one of the first crops to be easily cultivated on a large scale, and had the added advantage of yielding a harvest that could be stored for a long time. Wheat covers more land area than any other commercial crop, and is the most important staple food for humans.

Maize, or corn, was commonly grown throughout the Americas in the late fifteenth and early sixteenth centuries. Explorers and traders carried maize back to Europe and introduced it to other countries. It then spread to the rest of the world, owing to its ability to grow in different environments. Sugar-rich varieties called sweet corn are usually grown for human consumption, while field corn varieties are used for animal feed and **biofuel**. Maize is the most widely grown grain crop in the Americas, with 361 million metric tonnes grown annually in the United States alone.

FIGURE 2 Wheat is used in a wide variety of foods such as breads, biscuits, cakes, breakfast cereals and pasta.

TABLE 1 Top 10 maize producers, 2015

Country	Production (million tonnes)
United States	363
China	229
Brazil	77
European Union	65
Ukraine	26
Argentina	25
India	23
Mexico	23
South Africa	13
Russia	12

Source: http://www.perfectinsider.com/top-ten-maize-producing-countries-in-the-world/

FIGURE 3 Corn cobs drying outside in Serbia

Fish is a staple food in some societies. The oceans provide an irreplaceable, renewable source of food and nutrition essential to good health. According to the United Nations Food and Agriculture Organization, about 75 per cent of fish caught is used for human consumption. The remainder is converted into fishmeal and oil, used mainly for animal feed and farmed fish.

In general, people in developing countries, especially those in coastal areas, are much more dependent on fish as a staple food than those in the developed world. About 1 billion people rely on fish as their primary source of animal protein.

Use the **United Nations Food and Agriculture Organization: Fisheries** weblink in the Resources tab to find out what is being done to promote sustainable aquatic biomes.

FIGURE 4 A fish haul in Bali, Indonesia

3.3 Activities

To answer questions online and to receive **immediate feedback** and **sample responses** for every question, go to your learnON title at www.jacplus.com.au. *Note*: Question numbers may vary slightly.

Remember

1. Make a list of the main staple foods of the world and the *places* (continents) where they are grown.
2. What is biofuel?

Explain

3. Explain why plants, rather than animals, dominate as the major staple foods of the world.
4. Australia is a major exporter of wheat. Why is Australia able to produce such a surplus?

Predict

5. With the increase in world population and greater pressure on fish stocks, what could be done to sustain fish stocks in oceans and lakes?
6. Maize is currently used as a feed for animals, as biofuel and as food for humans. Why might this be an *unsustainable environmental* practice in future?

Think

7. Although fish may be seen as a staple food for many people, why is it not possible for fish to be a staple food for everyone?
8. Referring to table 1, why do you think countries other than those in the Americas are producing large quantities of maize?

my **World** Atlas | **Deepen your understanding of this topic with related case studies and questions.**
❍ **Rice**
❍ **Wheat**

3.4 How does traditional agriculture produce food?

3.4.1 Subsistence agriculture

In the more developed countries of the world, people have easy access to food in stores such as supermarkets. However, many people in developing nations are still tied to **subsistence** agriculture and visits to local markets to buy and exchange food.

For many people, changes in technology and the development of sophisticated agricultural practices have, over millennia, removed the need to hunt and gather food. With modern forms of transport, there is now enormous movement of food stocks around the world. Nevertheless, many people around the world still practise traditional agriculture.

FIGURE 1 World agricultural practices and food production

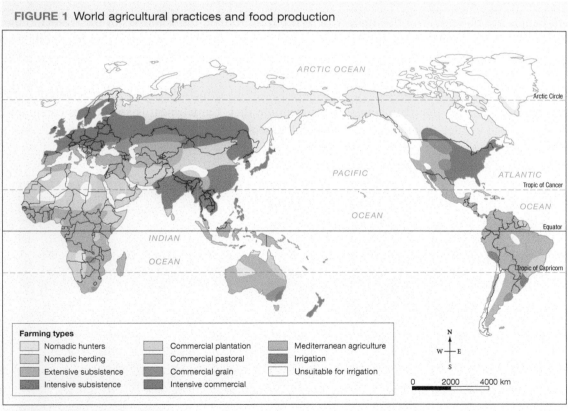

Source: Reproduced from www.S-cool.co.uk

3.4.2 Hunters and gatherers: the San people

Today, about 50000 San people (or Kalahari Bushmen) live in the Kalahari Desert in southern Africa. Approximately 6 per cent still live in the traditional way.

Traditionally **nomadic** San people travel in small family groups, roaming over regions of up to 1000 square kilometres. They have no pack animals, and carry few possessions — only spears, bows and arrows, bowls and water bags. The bushmen's San people's clothes are made from animal skins. When needed, they construct dome-shaped shelters of sticks that are thatched with grass.

The San people are experts at finding water and tracking animals. The men hunt antelope and wildebeest, while the women hunt small game such as lizards, frogs and tortoises, and gather roots, berries and grubs. When the waterholes are full, empty ostrich shells are filled with water, and buried in the sand for times of drought.

FIGURE 2 A San tribesman teaches his son how to use a bow and arrow.

3.4.3 Nomadic herders: the Bedouin people

Bedouin people are nomads who live mainly in Syria, Iraq, Jordan, the countries of the Arabian Peninsula, and the Sahara. Some groups are camel herders who live in the inner desert regions. Others herd sheep and goats on the desert fringes, where more water is available. Unless Bedouin communities find a good piece of grazing land, they rarely stay in one place longer than a week.

Bedouin camel-herding families can survive on as few as 15 camels. The camels provide not only transportation but also milk — the main staple of the Bedouin diet. Camel meat is sometimes eaten, and dried camel dung is used as fuel. Camel hair is collected and woven into rugs and tent cloth.

FIGURE 3 A Bedouin camp in Saudi Arabia

3.4.4 Shifting agriculture: the Huli people

The Huli people live in the rainforests of the Papua New Guinea highlands. Many still lead a traditional way of life. The land on which they live has steep hillsides and dense rainforest.

The Huli people today use a farming system known as shifting agriculture. This means that land is used for food production until its fertility declines; it is then abandoned until its fertility returns naturally. The Huli people clear a patch of rainforest and plant crops of sweet potato, sugar cane, corn, taro and green vegetables. It is the role of the women to tend these gardens, and their individual huts are built next to the gardens. The men live together in a communal house.

FIGURE 4 A map showing Huli land, Papua New Guinea

Source: MAPgraphics Pty Ltd, Brisbane

When the soil of the garden no longer produces good crops, a new patch of rainforest is cleared, leaving the old one to recover naturally. The garden crops are supplemented by food that the men have hunted. Wild and domesticated pigs are a common source of meat.

While most Huli people still live in their traditional lands, they wear some items of Western-style clothing, and knives, cooking utensils and mirrors are common.

FIGURE 5 A Huli tribesman, Papua New Guinea

3.4 Activities

To answer questions online and to receive **immediate feedback** and **sample responses** for every question, go to your learnON title at www.jacplus.com.au. *Note*: Question numbers may vary slightly.

Remember

1. Name three major types of traditional food production.
2. In which *places* in the world would you find Bedouin people?

Explain

3. Describe the traditional lifestyle of the San people.
4. Explain the advantages and disadvantages of nomadic herding practices regarding *sustainability*.

Discover

5. Why do people who live in rainforests need to use their *environment's* resources *sustainably*?
6. Name one impact on people and one impact on the natural *environment* if rainforests in the world were to *change* by decreasing significantly.

Predict

7. How might modern technology affect the Huli people in the next 25 years with respect to their traditions and food production practices?
8. What *changes* may occur to the way of life of nomadic herders in the future?

Think

9. In what ways are the traditional lives of hunters and gatherers and nomadic herders similar and different?
10. Referring to figure 1, explain why food production is concentrated in the *places* and *spaces* bordering the tropical zones.

3.5 How did Indigenous Australian peoples achieve food security?

3.5.1 Introduction

Since the beginning of the Dreaming Aboriginal and Torres Strait Islander peoples established management practices of their lands, waterways, lakes and marine environments to ensure food security. At the time of European occupation in 1788 most Aboriginal and Torres Strait Islander peoples were hunters and gatherers. However, some nations had abundant food supplies in their regions and were able to settle in one place. In all cases their deep knowledge and close association with the land allowed for sustainable management of the ecosystems and biomes in which they lived. The 'world view' that describes this sustainable lifestyle is called an 'earth-centred' approach. This means people's interaction with the environment is one of a caring stewardship.

3.5.2 Aboriginal and Torres Strait Islander peoples' food sourcing

Aboriginal and Torres Strait Islander peoples sourced their foods from a wide range of uncultivated plants and wild animals, with some estimates suggesting there were up to 7000 different sources of food. The composition of the food was greatly influenced by both the season and geographic location of the community region.

In Aboriginal and Torres Strait Islander communities there was a division of labour among men, women and children. Food sources based on cereals, fruits and vegetables were collected or gathered daily by women and children. Men were involved more in hunting for game and fishing, as well as wider scale land management using fire.

To ensure food security, communities developed a range of food-gathering techniques that were sustainable. For example, some seeds from gathered plants were left behind to allow for new growth, and a few eggs were always left in nests to hatch. This ensured that species would survive and communities could expect to find food in the same place in the future. See figure 1 for details of food types from both tropical and temperate regions of Australia, including arid and desert regions.

FIGURE 1 A generalised selection of different foods and water resources

What did Aboriginal and Torres Strait Islander peoples use to achieve food (and water) security?

Water: Water was obtained from rivers, lakes, rock holes, soaks, beds of intermittent creeks and dew deposited on surfaces. Moisture obtained from foods such as tree roots and leaves also provided water.

Cereal foods: Grass seeds from the clover fern were ground to form flour for damper. Many other seed types were similarly treated.

Fruit and vegetables: Fruits, berries, orchids and pods were available, depending on the region and seasonal availability (for example, sow thistle, lilly pilly, pigface fruit, kangaroo apple, wild raspberry, quandong and native cherry) as well as wild figs, plums, grapes and gooseberries. Also eaten were plant roots such as bull rushes, yams and bulbs; the heart of the tree fern and the pith of the grass tree; and the blister gum from wattles, native truffles and mushrooms.

Eggs: Emu, duck, pelican and many other birds' eggs were eaten.

Meat: Meats included insects such as the larval stage of the cossid moth or witchetty grub and the Bogong moth, honey ants, native bees and their honey and scale insects; animals such as kangaroos, emus, eels, crocodiles, sea turtles, snakes, goannas and other lizards; and birds such as ducks, gulls and pelicans.

Fish and shellfish: Freshwater fish such as perch, yabbies and mussels in creeks and rock holes as well as saltwater fish of all varieties were caught.

Medicines: Over 120 native plants were used as sedatives, ointments, diarrhoea remedies, and cough and cold palliatives as well as for many other known treatments.

3.5.3 Torres Strait Islander people

Torres Strait Islander peoples' food sources, both historically and today, are based on fishing, horticulture and inter-island trading activities. Torres Strait Islander peoples have a profound understanding of the sea, including its tides and sea life. While their food sources vary from island to island, their lifestyle can be best described as subsistence agriculture with seafood, garden foods and other produce stored and preserved for both local use and trade.

3.5.4 The use of fire

The use of fire was a significant aspect of the Indigenous Australian peoples' land management. What has been described as the 'park-like' landscape of the Australian bush was purposely created by clearing forest in a controlled burn using fire sticks. After the fires, new plant growth with tender shoots attracted all types of birds and animals to the area. The grassland areas that resulted from the controlled burning of the landscape became ideal places to hunt kangaroos. Burning also caused animals to be flushed out into the open where they could be speared (see figure 3). The Indigenous Australian peoples' use of fire had to be carefully managed as part of their efforts to ensure food security and as such was a sustainable practice based on a sound knowledge of fire control.

FIGURE 2 Cooking bush food in a traditional Kup Murrie, or ground oven

FIGURE 3 Using fire for hunting and to manage the land

3.5.5 The arrival of Europeans

Since European occupation, Aboriginal and Torres Strait Islander peoples have been displaced from their lands and traditional methods of sourcing food have mostly ceased. This is because government policies forced Indigenous communities to move to missions and as a consequence there was a denial of access to their lands. However, some methods of food sourcing continued up until recent times in areas the Europeans hadn't yet occupied, including in remote arid and desert areas and in the sparsely settled northern parts of Australia.

Today, many Aboriginal and Torres Strait Islander peoples, particularly in remote areas, are suffering from food insecurity due to the forced move away from nutritious bush tucker and historical government policies. This, among other things, has led to health issues such as low life expectancy, and related issues such as poor education and reduced work opportunities.

More recently, with the introduction of the Native Title Act (1993) and involvement of state governments, members of Indigenous communites have been able to re-establish connections with their lands through collaborative land and water management projects.

3.5.6 The Bogong moth: a past food source

While there were many other sources of food for Indigenous communities that lived near the south-eastern Australian highlands, the Bogong moth was a particularly important seasonal speciality. The Bogong moth, which lived in the ground as larvae in Queensland, migrated in millions to the south-eastern highlands to seek out cool, rocky overhangs and crevices where they could sleep through the long, hot summer months, surviving off the fat in their bodies (see figure 4).

The Bogong moths were a rich source of fat and protein for Indigenous Australian peoples who lived adjacent to the highlands of Victoria and New South Wales. Many culture groups would migrate from the valleys and foothills into the highlands and set up camps for the feasting ceremony. They would smoke out the moths, which they called 'cori', collecting them by the thousands to be cooked over hot rocks for feasting. In addition to savouring this important seasonal food source, making the annual pilgrimage to the high country presented these groups with an important opportunity to interact socially, participate in ceremonies and to arrange inter-community marriages.

FIGURE 4 Massed Bogong moths on a rock face

3.5.7 Eel farming by the Gunditjmara people

The home of the Gunditjmara people, the Budj Bim National Heritage Landscape, is the site of one of Australia's largest ancient aquaculture systems. This area, which is part of the Mount Eccles National Park near Portland in Victoria, shows evidence of a large, permanent settlement of stone huts and channels used for farming and the local trade of eels (see figure 5). The Gunditjmara people managed this landscape by digging channels and constructing weirs to bring water and young eels from Darlot Creek to local ponds and wetlands. Woven baskets placed at the weirs were used to harvest the mature eels. The area provided an abundance of food, ensuring food security for all.

FIGURE 5 Remains of Aboriginal stone eel traps at Lake Condah

Following European occupation of the area in the 1830s, the Gunditjmara people fought for their lands in the Eumerella Wars, which lasted for more than 20 years. By the 1860s the remaining Gunditjmara people were displaced to a government mission at Lake Condah. The mission lands were returned to the Gunditjmara people in 1987 and thereafter the Deen Maar Indigenous Protected Area (IPA) was declared in 1999 and the area was listed on the Australian National Heritage register in 2004.

Today the Gunditjmara people, as part of the Winda-Mara Aboriginal Corporation, manage the 248-hectare Darlots Creek (Killara), which flows from Lake Condah in the Budj Bim National Heritage Landscape. They aim to reinstate the wetlands and manna gum woodlands and re-establish the eel aquaculture industry as a sustainable business prospect. Further works are in progress to control weeds and feral animals and to expand tourism by building visitor boardwalks and information signage.

3.5 Activities

To answer questions online and to receive **immediate feedback** and **sample responses** for every question, go to your learnON title at www.jacplus.com.au. *Note*: Question numbers may vary slightly.

Remember
1. What was the division of labour for men, women and children when sourcing food?
2. How did Indigenous Australian peoples use fire to source food?
3. Who are the traditional owners of the Budj Bim region of Victoria?

Explain
4. Why did Aboriginal and Torres Strait Islander peoples lose access to 'bush tucker' after European occupation?
5. How did the Indigenous Australian peoples access the Bogong moth as a food source?

Predict
6. How might the re-establishment of eel farming help the economic opportunities of the Gunditjmara people?

Think
7. What lessons could be learned from the way the Budj Bim National Heritage Landscape is managed and applied to a national park with which you are familiar?
8. Which foods that Aboriginal and Torres Strait Islander peoples sourced could be utilised in cooking today? Have you tried any of the wide range of traditional food types? If so, how tasty or appetite satisfying were they to you?

3.6 SkillBuilder: Constructing ternary graphs

WHAT ARE TERNARY GRAPHS?

Ternary graphs are triangular graphs that show the relationship or interconnection between three features. They are particularly useful when a feature has three components and the three components add up to 100 per cent. Ternary graphs are most often used to show soil types, employment structures and age structures.

FIGURE 1 Economic activity in selected countries

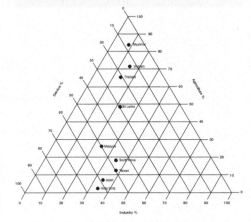

Go online to access:

- a clear step-by-step explanation to help you master the skill
- a model of what you are aiming for
- a checklist of key aspects of the skill
- a series of questions to help you apply the skill and to check your understanding.

learn on RESOURCES — ONLINE ONLY

Watch this eLesson: Constructing ternary graphs (eles-1728)

Try out this interactivity: Constructing ternary graphs (int-3346)

3.7 How have we increased our food?

3.7.1 How does food increase come about?

World food production has grown substantially over the past century. Increased fertiliser application and more water usage through irrigation have been responsible for over 70 per cent of crop yield increases. The Second Agricultural Revolution in developed countries after World War II, and the **Green Revolution** in developing countries in the mid 1960s, transformed agricultural practices and raised crop yields dramatically.

Since the 1960s agriculture has been more productive, with world per capita agricultural production increasing by 25 per cent in response to a doubling of the world population.

3.7.2 Environmental factors

In the past, growth in food production resulted mainly from increased crop yields per unit of land and to a lesser extent from expansion of cropland. From the early 1960s, total world cropland increased by only 9 per cent, but total agricultural production grew nearly 60 per cent. Increases in yields of crops, such as sweet potatoes and cereals, were brought about by a combination of:

- increased agricultural inputs
- more intensive use of land
- the spread of improved crop varieties.

In some places, such as parts of Africa and South-East Asia, increases in fisheries (areas where boats are used to catch fish) and expansion of cropland areas were the main reasons for the increase in food supply.

In addition, cattle herds became larger. In many regions — such as in the savanna grasslands of Africa, the Andes, and the mountains of Central Asia — livestock is a primary factor in food security today. Fertilisers have increased agricultural outputs and enabled more intensive use of the land. Global fertiliser use is likely to rise to above 200.5 million tonnes in 2018, 25 per cent higher than recorded in 2008.

TABLE 1 Fertiliser use, 1959–60, 1989–90 and 2020

Region/nutrient	Fertiliser use			Annual growth	
	1959/60	1989/90	2020	1960–90	1990–2020
	(million nutrient tonnes)			(per cent)	
Developed countries	24.7	81.3	86.4	4.0	0.2
Developing countries	2.7	62.3	121.6	10.5	2.2
East Asia	1.2	31.4	55.7	10.9	1.9
South Asia	0.4	14.8	33.8	12.0	2.8
West Asia/North Africa	0.3	6.7	11.7	10.4	1.9
Latin America	0.7	8.2	16.2	8.2	2.3
Sub-Saharan Africa	0.1	1.2	4.2	8.3	3.3
World total	27.4	143.6	208.0	5.5	1.2
Nitrogen	9.5	79.2	115.3	7.1	1.3
Phosphate	9.7	37.5	56.0	4.5	1.3
Potash	8.1	26.9	36.7	4.0	1.0

Sources: Bumb, B. and C. Baanante. 1996. World Trends in Fertilizer Use and Projections to 2020. Policy Brief 38, Table 1. Washington, DC: International Food Policy Research Institute http://www.ifpri.org/publication/world-trends-fertilizer-use-and-projections-2020

3.7.3 Trade factors and economic factors

From the 1960s onwards, there has been significant growth of world trade in food and agriculture. Food imports to developing countries have grown, together with imports of fertilisers, thus reducing the likelihood of developing countries suffering from famine.

TABLE 2 Share of crop production increases, 1961–2030

	Arable land expansion (1)		Increases in cropping intensity (2)		Harvested land expansion (1+2)		Yield increases	
	1961–1999	1997/99–2030	1961–1999	1997/99–2030	1961–1999	1997/99–2030	1961–1999	1997/99–2030
All developing countries	23	21	6	12	29	33	71	67
South Asia	6	6	14	13	20	19	80	81
East Asia	26	5	–5	14	21	19	79	81
Near East/North Africa	14	13	14	19	28	32	72	68
Latin America and the Caribbean	46	33	–1	21	45	54	55	46
Sub-Saharan Africa	35	27	31	12	66	39	34	61
World	15		7		22		78	

3.7.4 What was the Green Revolution?

The Green Revolution was a result of the development and planting of new **hybrids** of rice and wheat, which saw greatly increased yields. There have been a number of green revolutions since the 1950s, including those in:

- the United States, Europe and Australia in the 1950s and 1960s
- New Zealand, Mexico and many Asian countries in the late 1960s, 1970s and 1980s.

With its high-yield varieties of cereals, chemical fertilisers and pesticides, and irrigation, the Green Revolution has had a very positive effect on global food production.

What happened?

The Green Revolution saw a rapid increase in the output of cereal crops — the main source of calories in developing countries. Farmers in Asia and Latin America widely adopted high-yielding varieties. Governments, especially in Asia, introduced policies that supported agricultural development. In the 2000s, cereal harvests in developing countries were triple those of 40 years earlier,

FIGURE 1 Spreading fertiliser in the Punjab, India, during the Green Revolution

while the population was a little over twice as large. Yield gains accounted for much of the increase in cereal output and calorie availability. Planting of these varieties coincided with expanded irrigation areas and fertiliser use, as seen in figure 1 where fertiliser is being spread in the Punjab.

3.7 Activities

To answer questions online and to receive **immediate feedback** and **sample responses** for every question, go to your learnON title at www.jacplus.com.au. *Note*: Question numbers may vary slightly.

Remember

1. In the past, what were the two reasons for the increase in food production?
2. Refer to table 1. Describe the trends in the use of fertilisers and irrigated land from 1965 to 2000.

Explain

3. Explain the significance of trade in food production.
4. Discuss the three reasons for improved crop production.

Discover

5. Research the background of the Green Revolution — why it occurred and the key *places* involved.
6. Investigate the *changes* that came about as a result of the Green Revolution.

Predict

7. Some scientists are suggesting that there will be a new Green Revolution. Investigate current thinking and predict the potential *scale* of this possible agricultural *change*.

Think

8. Were the *changes* brought about during the Green Revolution successful? When explaining your decision, refer to the Punjab and other *places*.

3.8 How are biomes modified for agriculture?

3.8.1 How do we use technology for food production?

In the twentieth century, rapid global population growth gave rise to serious concerns about the ability of agriculture to feed humanity. However, additional gains to food production have come from newer processes and technology.

Across the world, humans have modified biomes to produce food through the application of innovative technologies. In general, the focus of agriculture is to modify water, climate, soils, land and crops.

3.8.2 How do we modify climate?

Irrigation is the artificial application of water to the land or soil to supplement natural rainfall. It is used to assist in the growing of agricultural crops to increase food production in dry areas and during periods of inadequate rainfall.

In flood irrigation, water is applied and distributed over the soil surface by gravity. It is by far the most common form of irrigation throughout the world, and has been practised in many areas, virtually unchanged, for thousands of years.

Modern irrigation methods include computer-controlled drip systems that deliver precise amounts of water to a plant's root zone.

Another way of modifying climate is with the use of greenhouses (or glasshouses) used for growing flowers, vegetables, fruits and tobacco (see figure 2). Greenhouses provide an artificial biotic environment to protect crops from heat and cold and to keep out pests. Light and temperature control allows greenhouses to turn non-arable land into arable land, thereby improving food production in marginal environments. Greenhouses allow crops to be grown throughout the year, making them especially important in high-latitude countries.

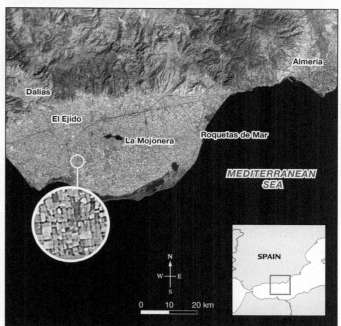

FIGURE 1 False-colour satellite image of greenhouses in the Almeria region

Source: American Geophysical Union and Google Maps

The largest expanse of plastic greenhouses in the world is around Almeria, in south-east Spain. Here, since the 1970s, semiarid pasture land has been replaced by greenhouse **horticulture** (see figures 1 and 2). Today, Almeria has become Europe's market garden. To grow food all year round, the region has about 26 000 hectares of greenhouses.

FIGURE 2 Inside an Almerian greenhouse

FIGURE 3 Land reclamation in the Netherlands

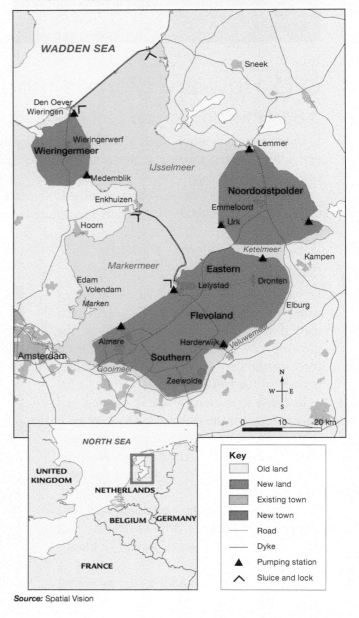

Source: Spatial Vision

3.8.3 How do we modify soils?

Fertilisers are organic or inorganic materials that are added to soils to supply one or more essential plant nutrients. Fertilisers are essential for high-yield harvests, and it is estimated that about 40 to 60 per cent of crop yields are due to fertiliser use. It is estimated that almost half the people on Earth are currently fed as a result of adding fertiliser to food crops.

3.8.4 How do we modify landscapes?

People change landscapes in order to produce food. **Undulating** land can be flattened, steep slopes terraced, or stepped, and wetlands drained. Land reclamation is the process of creating new land from seas, rivers or lakes. In addition, it can involve turning previously unfarmed land, or degraded land, into arable land by fixing major deficiencies in the soil's structure, drainage or fertility.

In the Netherlands, the Dutch have tackled huge reclamation schemes to add land area to their country. One such scheme is the IJsselmeer (see figure 3), where four large areas (*polders*) have been reclaimed from the sea, adding an extra 1650 square kilometres for cultivation. This has increased the food supply in the Netherlands and created an overspill town for Amsterdam.

3.8 Activities

To answer questions online and to receive **immediate feedback** and **sample responses** for every question, go to your learnON title at www.jacplus.com.au. *Note*: Question numbers may vary slightly.

Remember

1. What *changes* to the *environment* are made by land reclamation?

Explain

2. Refer to figure 3. Use the *scale* to calculate the approximate area of new land created in Flevoland.

Discover

3. How is land that is reclaimed from the sea, such as the Netherlands' polders, made productive for farming and food production?

Predict

4. Refer to figures 1 and 2. How do greenhouses modify *spaces* and *places* on the Earth's surface?
5. Refer to figure 3. What might be the purpose of the pumping stations?

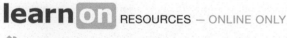 RESOURCES — ONLINE ONLY

🔹 **Try out this interactivity:** Changing nature (int-3321)

3.9 How is food produced in Australia?

3.9.1 Why are farms found in certain locations?

Modern food production in Australia can be described as commercial agricultural practices that produce food for local and global markets. Farms may produce single crops, such as sugar cane, or they may be mixed farms that produce sheep and cereals, for example. Farms use sophisticated technology, and in many cases are managed by large corporations with an **agribusiness** approach.

There is a wide range of types of agriculture in Australia, as shown in figure 1. These types occupy spaces across all biomes found in Australia, from the tropics to the temperate zones.

The location of farms in Australia shows that there is a change in the pattern of farming types, from the well-watered urban coastal regions towards the arid interior. Because much of Australia's inland rainfall is less than 250 millimetres, farm types in these places are limited to open-range cattle and sheep farming.

The pattern of land use and transition of farm types is shown in figure 2. It indicates that **intensive farms**, which produce perishables such as fruit and vegetables, are located on high-cost land close to urban markets. At the other extreme, the **extensive farms**, which manage cattle, sheep and cereals, are found on the less expensive lands distant from the market.

FIGURE 1 Types of agriculture in Australia

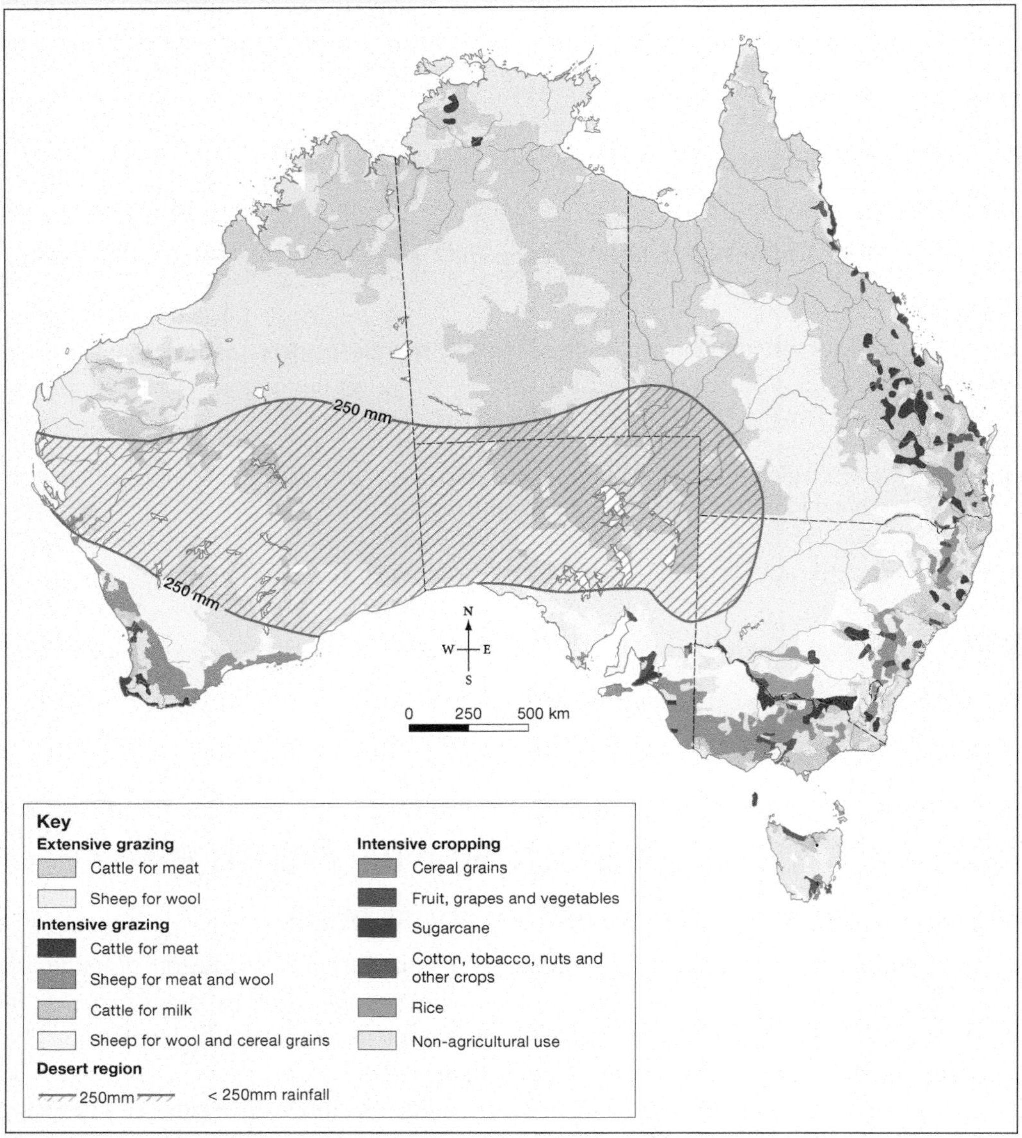

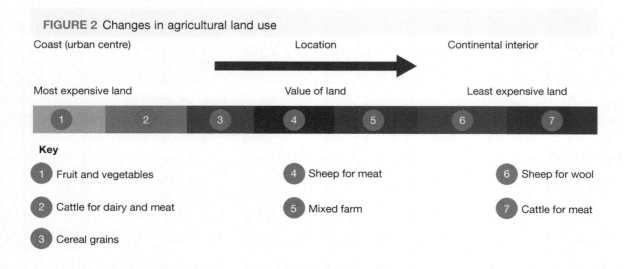

250 mm
250 mm
N
W
E
S
0 250 500 km
Key
Extensive grazing
Cattle for meat
Sheep for wool
Intensive grazing
Cattle for meat
Sheep for meat and wool
Cattle for milk
Sheep for wool and cereal grains
Desert region
250mm
< 250mm rainfall
Intensive cropping
Cereal grains
Fruit, grapes and vegetables
Sugarcane
Cotton, tobacco, nuts and other crops
Rice
Non-agricultural use

Source: © Commonwealth of Australia (Geoscience Australia) 2013

FIGURE 2 Changes in agricultural land use

Coast (urban centre)
Location
Continental interior
Most expensive land
Value of land
Least expensive land
1
2
3
4
5
6
7
Key
1 Fruit and vegetables
2 Cattle for dairy and meat
3 Cereal grains
4 Sheep for meat
5 Mixed farm
6 Sheep for wool
7 Cattle for meat

3.9.2 Some farm types in Australia

Extensive farming of sheep or cattle

Sometimes known as livestock farming or grazing, sheep and cattle stations are found in semi-arid and desert grassland biomes, with rainfall of less than 250 millimetres. In 2015 Australia's 70 million sheep and 27 million cattle were found mainly in Queensland, Victoria and New South Wales. Farms are large in scale, covering hundreds of square kilometres. These days, they have very few employees, and often use helicopters and motor vehicles for mustering. Meat and wool products go to both local and overseas markets for cash returns.

Wheat farms

About 25 000 farms in Australia grow wheat as a major crop, and the average farm size is 910 hectares, or just over 9 square kilometres. Wheat production in Australia for 2015 was estimated at 23 million tonnes. As in other areas of the world, extensive wheat farming is found in mid latitude temperate climates that have warm summers and cool winters, and annual rainfall of approximately 500 mm. In Australia, these conditions occur away from the coast in the semi-arid zone. The biome associated with this form of food production is generally open grassland, **mallee** or savanna that has been cleared for the planting of crops.

Soils can be improved by the application of fertilisers, and crop yields increased by the use of disease-resistant, fast-growing seed varieties. Wheat farms are highly mechanised, using large machinery for ploughing, planting and harvesting. The farm produce, which can amount to 2 tonnes per hectare, is sold to large corporations in local and international markets.

FIGURE 3 Cattle mustering

Mixed farms

Mixed farms combine both grazing and cropping practices. They are located closer to markets in the wetter areas, and are generally small in scale, but operate in much the same way as cattle and sheep farms.

Intensive farming

Intensive farms are close to urban centres, producing dairy, horticulture and market gardening crops. They produce milk, fruit, vegetables and flowers, all of which

FIGURE 4 Wheat farming with a combine harvester

are perishable, sometimes bulky, and expensive to transport. The market gardens are capital- and labour-intensive, because the cost of land near the city is high, and many workers are required for harvesting.

Plantation farming

This form of agriculture is often found in warm, well-watered tropical places. Plantations produce a wide range of produce such as coffee, sugar cane, cocoa, bananas, rubber, tobacco and palm oil. Farm sizes can be 50 hectares or more in size. Although many such farms in Australia are family owned, in other parts of the world they are often operated by large multinational companies. Biomes that contain plantations are mainly tropical forests or savanna, and require large-scale clearing to allow for farming. Cash returns are high, and markets are both local and global.

FIGURE 5 Strawberries are typically grown in market gardens.

FIGURE 6 A banana plantation near Carnarvon, Western Australia

3.9 Activities

To answer questions online and to receive **immediate feedback** and **sample responses** for every question, go to your learnON title at www.jacplus.com.au. *Note*: Question numbers may vary slightly.

Remember

1. Which type of agricultural land use is closest to urban centres, and which is the furthest away?
2. How does the *environment* in the centre of Australia affect farming types?
3. What is the *interconnection* between climate and farm type in Australia? (*Hint:* Refer to a climate map in your atlas for other ideas.)

Explain

4. Explain why extensive, large-*scale* cattle and sheep farms are typically located in remote and arid regions of Australia.

5. Using the map of farm production in Australia, describe and explain the location of:
   (a) wheat farms
   (b) dairy farms.

Discover

6. Investigate which foods are grown closest to you.
7. Collect information on the percentage of land used for the different forms of farming in Australia, and then show this data in a graph. Comment on the details shown in your graph.
8. One of the growing plantation industries is that of palm oil. It often has great impacts on tropical biomes; loss of habitat is one such impact. On a world map, locate major palm oil production areas and explain the implications of loss of habitat in those areas.
9. Various plantations in Queensland (such as pineapple, sugar cane and banana plantations) are associated with fertiliser run-off, which is affecting the Great Barrier Reef. Find out what effects fertiliser has on these aquatic *environments*.

Predict

10. What would be the impact of flood or drought on any of the commercial methods of food production?
11. Predict the impact of the growth of Australian capital cities on the *sustainability* of surrounding market gardens.

Think

12. Why is much of Australia's food production available for export?
13. It used to be said that Australia's economy 'rode on the sheep's back'. What do you think this means, and do you think it is still true today?

3.10 What does a farming area look like?

Modern-day food production relies heavily on technology to create ideal farming conditions. This may involve reshaping the land to allow for large agricultural machinery and for the even distribution and drainage of water. Uneven or unreliable rainfall can be supplemented by irrigation. As a result of such changes, large areas can become important farmland.

CASE STUDY

Griffith

Griffith, located in the Western Riverina of New South Wales, is an important agricultural and food-processing centre for the region, generating more than $1.9 billion dollars' worth of food. It is responsible for 60 per cent of the oranges, 44 per cent of the rice and 51 per cent of the wine produced in New South Wales.

The first European explorer to the area was John Oxley, who described the region as 'uninhabitable and useless to civilised man'. This was largely due to the lack of a suitable water supply. The construction of irrigation canals in 1906 established a reliable source of water that could be used in food production. The region has become an important food centre owing to the large-scale use of irrigation combined with suitable flat land, fertile soils and a mild climate.

To investigate the area in a little more detail, study the topographic map shown in figure 2.

FIGURE 1 Orange trees growing in an orchard near Griffith, New South Wales

FIGURE 2 Topographic map extract, Griffith, New South Wales

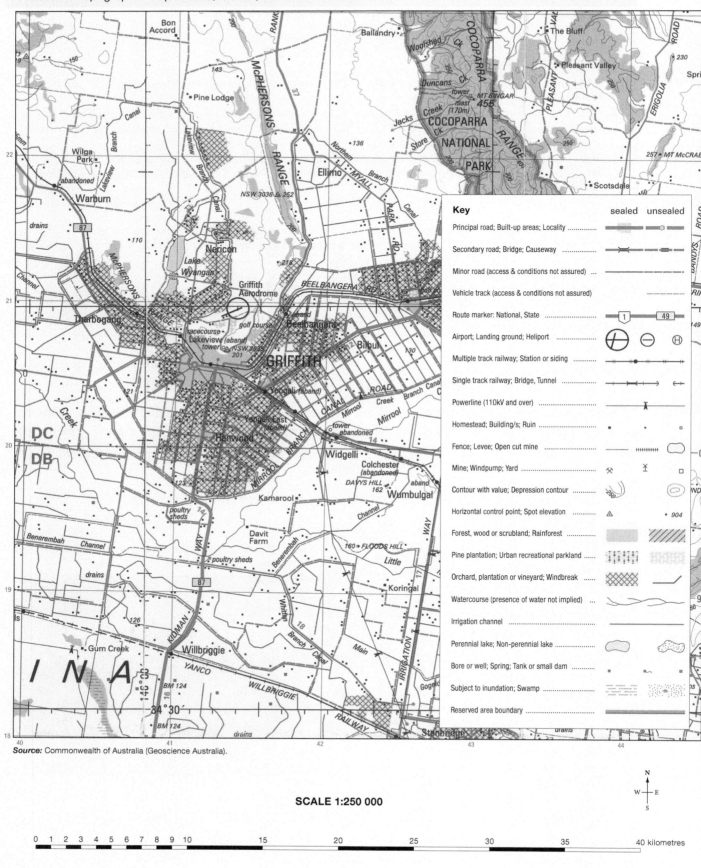

Source: Commonwealth of Australia (Geoscience Australia).

SCALE 1:250 000

3.10 Activities

To answer questions online and to receive **immediate feedback** and **sample responses** for every question, go to your learnON title at www.jacplus.com.au. *Note*: Question numbers may vary slightly.

Discover

1. What types of *environment* might have existed in the Griffith area when Oxley first arrived?
2. Identify and name a possible source for irrigation water on the map in figure 2.
3. How is water moved around this area? (*Hint:* Follow the blue lines.)
4. Using the contour lines and spot heights in figure 2 as a guide, estimate the average elevation of the map area.
5. What is the importance of topography (the shape of the land) to irrigation?
6. What types of farming are found at the following *places* in figure 2?
   (a) GR410195
   (b) GR413220
7. Approximately what percentage of the map in figure 2 area is irrigated?

Explain

8. Are orchards and vineyards an example of intensive or extensive farming? Explain.
9. Compare the pattern of irrigation channels and buildings in AR3919 and AR4220 in figure 2. Suggest a reason for the differences you can see.

Think

10. Within Griffith there are many factories that process raw materials, such as rice mills, wineries and juice factories. What would be the advantages and disadvantages of locating processing factories close to growing areas?

3.11 SkillBuilder: Describing patterns and correlations on a topographic map

WHAT ARE PATTERNS AND CORRELATIONS ON A TOPOGRAPHIC MAP?

A pattern is the way in which features are distributed or spread. A correlation shows how two or more features are interconnected—that is, the relationship between the features. Patterns and correlations in a topographic map can show us cause-and-effect connections.

Go online to access:

- a clear step-by-step explanation to help you master the skill
- a model of what you are aiming for
- a checklist of key aspects of the skill
- a series of questions to help you apply the skill and to check your understanding.

FIGURE 1
Topographic map extract showing the Clare Valley, South Australia

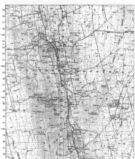

3.12 Why is rice an important food crop?

3.12.1 The importance of rice

Rice is the seed of a semi-aquatic grass. In warm climates, in more than 100 countries, it is cultivated extensively for its edible grain. Rice is one of the most important staple foods of more than half of the world's population, and it influences the livelihoods and economies of several billion people. In Asia, rice provides about 49 per cent of the calories and 39 per cent of the protein in people's diet. In 2010, approximately 154 million hectares of rice were harvested worldwide, and 95 kilograms were produced for each person on Earth.

Figure 1 shows that the largest concentration of rice is grown in Asia. About 132 million hectares are cultivated with this crop, producing 88 per cent of the world's rice. Of this, 48 million hectares and 31 per cent of the global rice crop are in South-East Asia alone.

Countries with the largest areas under rice cultivation are India, China, Indonesia, Bangladesh, Thailand, Vietnam, Myanmar (Burma) and the Philippines, with 80 per cent of the total rice area.

FIGURE 1 Top 10 rice-producing countries

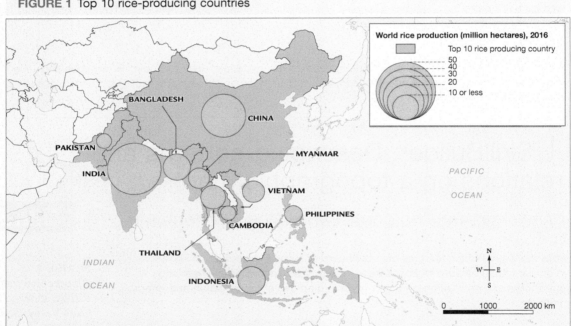

Source: WorldAtlas.com, http://www.worldatlas.com/articles/the-countries-producing-the-most-rice-in-the-world.html

3.12.2 Factors affecting rice production in Asia

Climate and topography

Rice can be grown in a range of environments that are hot or cool, wet or dry. It can be grown at sea level on coastal plains and at high altitudes in the Himalayas. However, ideal conditions in South-East Asia are high temperatures, large amounts of water, flat land and fertile soil.

In Yunnan Province, China, the mountain slopes have been cultivated in terraced rice paddies by the Hani people for at least 1300 years (see figure 2). The terraces stop erosion and surface run-off.

FIGURE 2 Spectacular rice terraces in Yunnan Province, China. These terraces are at an elevation of 1570 metres.

Irrigation

Traditional rice cultivation involves flooding the paddy fields (*padi* meaning 'rice plant' in Malay) for part of the year. These fields are small, and earth embankments (*bunds*) surround them. Rice farmers usually plant the seeds first in little seedbeds and transfer them into flooded paddy fields, which are already ploughed. Canals carry water to and from the fields. Houses and settlements are often located on embankments or raised islands near the rice fields.

Approximately 45 per cent of the rice area in South-East Asia is irrigated, with the largest areas being found in Indonesia, Vietnam, the Philippines and Thailand. High-yielding areas of irrigated rice can also be found in China, Japan and the Republic of Korea. Because water is available for most of the year in these places, farmers can grow rice all year long. This intensive scale of farming can produce two and sometimes three crops a year.

Upland rice is grown where there is not enough moisture to nurture the crops; an example of such cultivation takes place in Laos. This method produces fewer rice varieties, since only a small amount of nutrients are available compared to rice grown in paddy fields.

Pests and diseases

Rice yields can be limited if any of the following conditions exist:
- poor production management
- losses caused by weeds (biotic factor)
- pests and diseases (biotic factor)
- inadequate land formation and irrigation water
- inadequate drainage that leads to a build-up of salinity and alkalinity.

FIGURE 3 Planting rice in paddy fields in north-east Thailand

padi (rice plant)

bund (embankment)

FIGURE 4 Rice demonstration plots at the International Rice Research Institute in the Philippines

Technology

Agricultural biotechnology, especially in China, has produced rice that is resistant to pests. There are also genes for herbicide resistance, disease resistance, salt and drought tolerance, grain quality and photosynthetic efficiency. Genetic engineering may be the way of the future in rice cultivation in some parts of the world.

However, in the Philippines, a new strain of rice has been developed that grows well in soils lacking phosphorus. This could change crop yields considerably, and has been a result of cross breeding rather than genetic engineering.

Environmental issues

Increasing temperatures, due to global warming, may be causing a drop in rice production in Asia, where more than 90 per cent of the world's rice is produced and consumed. The Food and Agriculture Organization of the United Nations (FAO) has found that in six of Asia's most important rice-producing countries — China, India, Indonesia, the Philippines, Thailand and Vietnam — rising temperatures over the past 25 years have led to a 10–20 per cent decline in rice output.

Scientists state that if rice production methods cannot be changed, or if new rice strains able to withstand higher temperatures cannot be developed, there will be a loss in rice production over the next few decades as days and nights get hotter. People may need to turn to a new staple crop.

Rice growing is eco-friendly and has a positive impact on the environment. Rice fields create a wetland habitat for many species of birds, mammals and reptiles. Without rice farming, wetland environments created by flooded rice fields would be vastly reduced.

3.12.3 Factors affecting rice production in Australia

Climate and topography

Eighty per cent of rice produced in Australia consists of temperate varieties that suit climates with high summer temperatures and low humidity. Rice is grown in the Murrumbidgee valleys of New South Wales and the Murray valleys of New South Wales and Victoria. The scale of production is sophisticated.

Sowing and irrigation

In Australia, rice grows as an irrigated summer crop from September to March. Most of it is sown by aircraft rather than planted by hand. Experienced agricultural pilots use satellite guidance technology to broadcast seed accurately over the fields.

Before sowing, the seed is soaked for 24 hours and drained for 24 hours, leaving a tiny shoot visible on the seed. Once sown, it slowly settles in the soft mud, and within three to four days each plant develops a substantial root system and leaf shoot. After planting, fresh water is released from irrigation supply channels to flow across each paddy field until the rice plants are well established.

Most countries grow rice as a **monoculture**, whereas Australian rice grows as part of a unique farming system. Farmers use a **crop rotation** cycle across the whole farm over four to five years. This means that the growers have other agricultural enterprises on the farm as well as rice. This system, designed for efficiency, sustainability and safety, means Australian growers maintain water savings, and have increased soil nutrients, higher yields and much healthier crops.

Once Australian rice growers harvest their rice, they use the sub-soil moisture remaining in the soil to plant another crop — either a wheat crop or pasture for animals. This form of rotation is the most efficient in natural resource use and agricultural terms.

Pests and diseases

Rice bays (areas contained by embankments — see figure 5) are treated with a chemical application, which prevents damage by pests and weeds. Without this treatment, crop losses would be extensive.

FIGURE 5 Murrumbidgee irrigation area rice fields

In the last 100 days before harvesting, the rice plant has no chemical applications, so that when it is harvested, it is virtually chemical free.

Technology

Most farms use laser-guided land levelling techniques to prepare the ground for production. This gives farmers precise control over the flow of water on and off the land. Such measurement strategies have contributed to a 60 per cent improvement in water efficiency. Most of the equipment used on rice farms is fitted with computer-aided devices, such as GPS (global positioning systems), CAD (computerised whole farm design), GIS (geographical information systems) and remote sensing. Australian rice growers are the most efficient and productive in the world.

FIGURE 6 Harvesting rice near Griffith, New South Wales

Environmental issues

The rice industry encourages biodiversity enhancement and greenhouse gas reduction strategies. Some farms in southern New South Wales are avoiding the use of chemical fertilisers and pesticides by converting farms to biodynamic practices, and have avoided salinity by planting red gums.

3.12 Activities

To answer questions online and to receive **immediate feedback** and **sample responses** for every question, go to your learnON title at www.jacplus.com.au. *Note:* Question numbers may vary slightly.

Remember

1. Refer to figure 1. Which countries produce most of the world's rice?
2. What is meant by the term *monoculture*?

Explain

3. Explain why *places* in Asia are ideally suited to rice growing.
4. Use the **Terraced rice** weblink in to explain how the terraced rice fields shown in figure 2 have been formed.
5. Explain the *environmental* issues that may affect future rice production.
6. Describe and explain the similarities and differences between the rice cultivation methods used in Asia and Australia.

Discover

7. Investigate two different rice-growing *places* in Asia and describe the reasons for the different *environments*.
8. Investigate an example of an Australian rice farm and outline its yearly rice-growing cycle.
9. Research the *interconnection* between rice growing and the Murray River for ensuring a *sustainable environment*.

Predict

10. Predict how technology will influence *changes* to rice cultivation in both Asia and Australia.

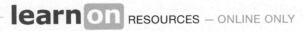

 myWorldAtlas   Deepen your understanding of this topic with related case studies and questions.
❍ **Rice**

3.13 Why is cacao a special food crop?

Access this subtopic at **www.jacplus.com.au**

3.14 Daly River: a sustainable ecosystem?

Access this subtopic at **www.jacplus.com.au**

3.15 How can aquaculture improve food security for Indigenous Australian peoples?

Access this subtopic at **www.jacplus.com.au**

3.16 Review

3.16.1 Review
The Review section contains a range of different questions and activities to help you revise and recall what you have learned, especially prior to a topic test.

3.16.2 Reflect
The Reflect section provides you with an opportunity to apply and extend your learning.

Access this subtopic at **www.jacplus.com.au**

TOPIC 4
What are the impacts of feeding our world?

4.1 Overview

Numerous **videos** and **interactivities** are embedded just where you need them, at the point of learning, in your learnON title at www.jacplus.com.au. They will help you to learn the content and concepts covered in this topic.

4.1.1 Introduction

The world's biomes support human life. We depend on these biomes for food, water, fibres, fuel and wood. We also rely on the services they provide, such as purifying air, regulating climate and providing spiritual and aesthetic comfort. However, our ecosystems are under considerable threat, largely as a result of our activities. Increasing population and demand for food have altered many of our biomes. There have been positive benefits, such as improved standards of living, although this has not been consistent across the globe. At the same time, there has been a significant decline in species biodiversity. Unless many of these problems are addressed, there is a growing concern that we may not be able to enjoy food security in future.

False-colour satellite images of the Wadi As-Sirhan Basin, Saudi Arabia, in 1987 (top) and 2012 (bottom). Large-scale irrigation in desert regions to maintain such agricultural fields consumes large quantities of groundwater.

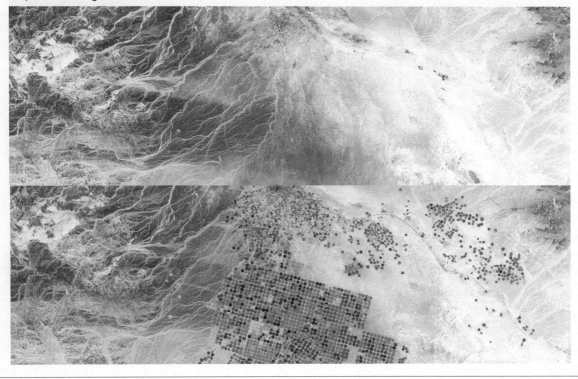

Starter questions

1. How can feeding the world be destroying the world?
2. Do you know where your food comes from?
3. What food items do you eat most of within a typical day? Are they animal-based, plant-based or fish-based?
4. Do you or your family grow any of your own food?
5. Looking at the images on the previous page, how might this irrigation farming affect the desert biome?

INQUIRY SEQUENCE

4.1	Overview	67
4.2	How does producing food affect biomes?	68
4.3	Where have all the trees gone?	70
4.4	**SkillBuilder:** GIS — deconstructing a map	online only 73
4.5	Paper profits, global losses?	online only 74
4.6	Should we farm fish?	74
4.7	**SkillBuilder:** Interpreting a geographical cartoon	online only 79
4.8	How do we lose land?	80
4.9	Irrigation: success or failure?	82
4.10	Does farming use too much water?	online only 84
4.11	Why is global biodiversity diminishing?	84
4.12	Does farming cause global warming?	online only 87
4.13	**Review**	online only 87

4.2 How does producing food affect biomes?

4.2.1 What is our biophysical world?

Food is essential to human life, and over the past centuries we have been able to produce more and more food to feed our growing population. While technology has enabled us to increase production, it has come at a price. Large-scale clearing of our forests, the overfishing of our oceans, and the constant overuse of soils has resulted in a significant decline in our biophysical world.

Planet Earth is made up of four spheres: the atmosphere, lithosphere, hydrosphere and biosphere (see figure 1).

All these spheres are interconnected and make up our biophysical or natural environment. For example, rain falling from a cloud (atmosphere) may soak into the soil (lithosphere) or flow into a river (hydrosphere) before being taken up by a plant or animal (biosphere) where it may be evaporated and returned to the atmosphere.

Natural events, such as storms or earthquakes, or human activities can create changes to one or all of these spheres. The production of food, whether from the land or sea, has the potential to change the natural environment and, in doing so, increases the likelihood of food insecurity. Table 1 shows how food production can affect the biophysical world.

FIGURE 1 The Earth's four spheres

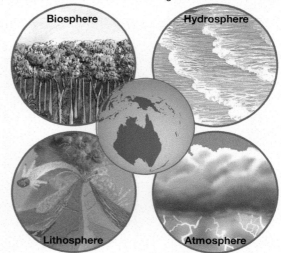

Hydrosphere: 97 per cent of the Earth's water is found in salty oceans, and the remainder as vapour in the atmosphere and as liquid in **groundwater**, lakes, rivers, glaciers and snowfields.

Biosphere: the collection of all Earth's life forms

Biosphere

Hydrosphere

Lithosphere

Atmosphere

Lithosphere: consists of the core, mantle and crust of the Earth

Atmosphere: contains all of the Earth's air

TABLE 1 How food production affects the biophysical world

Activities	Atmosphere	Lithosphere	Biosphere	Hydrosphere
Clearing of native vegetation for agriculture	x	x	x	x
Overgrazing animals		x	x	x
Overusing irrigation water, causing saline soils		x	x	x
Burning forests to clear land for cultivation	x	x	x	x
Run-off of pesticides and fertilisers into streams		x	x	x
Producing greenhouse gases by grazing animals and rice farming	x			
Changing from native vegetation to cropping		x	x	x
Withdrawing water from rivers and lakes for irrigation	x	x	x	x
Overcropping soils		x	x	x
Overfishing some species			x	

4.2.2 What has happened to our biophysical world?

Currently, the world produces enough food to feed all 7 billion people. We produce 17 per cent more food per person than was produced 30 years ago, and the rate of food production has been greater than the population growth. This has been the result of improved farming methods; the increased use of fertilisers and pesticides; large-scale irrigation; and the development of new technologies, ranging from farm machinery to better quality seeds.

There have been many benefits associated with this change, especially in terms of human wellbeing and economic development. However, at the same time, humans have changed the Earth's biomes more rapidly and more extensively than in any other time period. The loss of biodiversity and **degradation** of land and water (which are essential to agriculture) is not sustainable. With an expected population of 9.7 billion in 2050, it has been estimated that food production will need to increase by approximately 70 per cent. The global distribution of environmental risks associated with food production can be seen in figure 2.

FIGURE 2 State of the world's land and water resources for food and agriculture

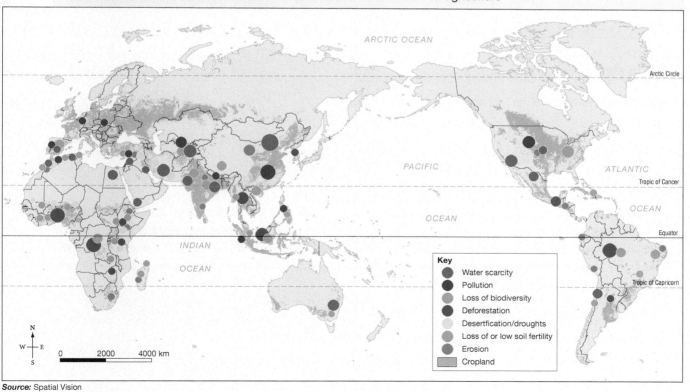

Source: Spatial Vision

4.2 Activities

To answer questions online and to receive **immediate feedback** and **sample responses** for every question, go to your learnON title at www.jacplus.com.au. *Note*: Question numbers may vary slightly.

Remember

1. Describe the **biophysical environment** of your local area.
2. Why has food production increased so rapidly over time?

Explain

3. Explain, with the use of figure 1 how a bird might *interconnect* with the four Earth spheres.
4. Select one example from table 1. Describe how human activity can *change* the biophysical world.
5. Refer to figure 2 and your atlas.
    (a) What are the main *environmental* issues facing Australia's food production?
    (b) In which *places* in the world is deforestation a major concern?
    (c) Which continents suffer from water scarcity?
    (d) What do you notice about the location and distribution of regions that do not have *environmental* problems relating to food production?

Discover

6. Select one agricultural product in Australia and conduct research to find data on how much is produced and how this has *changed* over time.

Think

7. Use the following labels to create a flow diagram showing how the clearing of native vegetation can affect all four of the Earth's spheres.
    − Soil left bare and exposed to wind and water erosion
    − Less evaporation of water from vegetation
    − Loss of habitat for birds, animals and insects
    − Increased water runs off from exposed land
    − Increased sediment builds up in streams
8. Refer to figure 2.
    (a) How would you rank these environmental impacts on a scale from the least to the most serious of impacts?
    (b) What criteria could you use to help make this decision?

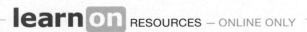 RESOURCES — ONLINE ONLY

 Try out this interactivity: Degrading our farmland (int-3323)

4.3 Where have all the trees gone?

4.3.1 Why are forests important?

Thirty per cent of the world's land surface is covered in forest. This is nearly one-third less than existed in prehistoric times, when humans did not have the technology to fell trees in large numbers. Large-scale deforestation has occurred as the need for timber products and land for food has increased over time. In the decade from 2000 to 2010, over 13 million hectares of forest was cleared each year. Cultivated land now covers one-quarter of the Earth's land surface.

Forest biomes offer us many goods and services, ranging from wood and food products, to supporting biological diversity. They provide habitat for a wide range of animals, plants and insects. Forests contribute to soil and water conservation, and they absorb **greenhouse gases**.

4.3.2 Why do we clear forests?

By clearing forests, valuable trees can be harvested for timber and paper production, while mining ores and minerals can be accessed below the Earth's surface. Sometimes, forests are flooded rather than cleared in order to construct dams for hydroelectricity. Forests may also be cleared for food production, such as small-scale subsistence farming, large-scale cattle grazing, and for **plantations** and crop cultivation. Figure 1 illustrates the main causes of deforestation in the Amazon rainforest during the peak period of deforestation. In recent years, the scale of deforestation in the Brazilian Amazon has been reduced due to an improved economy and the growing awareness of the value of preserving forests.

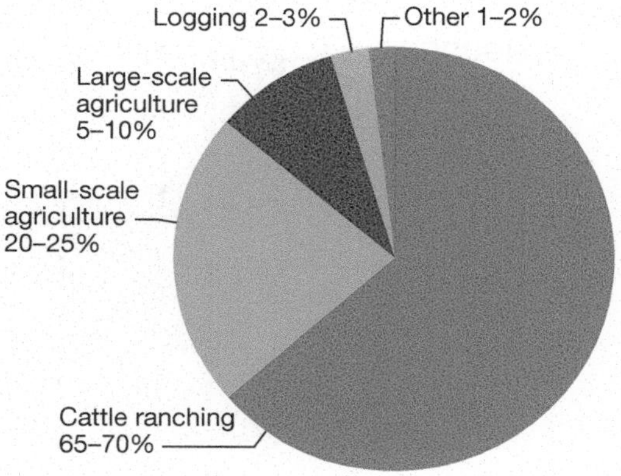

Road construction, usually funded by governments, also plays a part in changing rainforest environments (see figure 2). Roads help to improve access and make more land available, especially to the landless poor. They also reduce population pressures elsewhere by encouraging people to move to new places. At the same time, businesses benefit from improved access to mine resources and forest timbers, and are better able to establish large cattle ranches and farms.

FIGURE 1 Main causes of deforestation in the Brazilian Amazon, 2000–2005

Logging 2–3%
Other 1–2%
Large-scale agriculture 5–10%
Small-scale agriculture 20–25%
Cattle ranching 65–70%

FIGURE 2 The effects of road building in the Amazon. Settlements tend to follow a linear pattern along the roads and then gradually move inland, opening up the forests.

4.3.3 What happens when forests are cleared?

Figure 3 illustrates some changes that forest clearing in the Amazon can have on the environment.

FIGURE 3 Impacts of clearing the Amazon rainforest

A New farmland with mixed crops established

B Smoke from clearing and burning

C Newly cleared land, trees cut down and burned. This is called slash-and-burn agriculture.

D Weeds and exotic species invade edges of remaining forest

E New road gives access to more settlers and to animal poachers

F Large cattle ranch

G Introduced cattle erode the fragile topsoil with their hard hooves.

H Erosion of topsoil increases, caused by rain on exposed soils.

I Flooding increases as stream channel is clogged with sediment

J The river carries more sediment as soil is washed into streams.

K Fences stop movement of rainforest animals in search of food.

L Pesticides and fertilisers wash into river

M Farm is abandoned as soil fertility is lost

N Weeds and other species dominate bare land

O Harvesting of timber reduces forest biodiversity.

4.3 Activities

To answer questions online and to receive **immediate feedback** and **sample responses** for every question, go to your learnON title at www.jacplus.com.au. *Note:* Question numbers may vary slightly.

Remember

1. Refer to figure 1. What are the three main causes of deforestation in the Amazon?
2. What are the advantages and disadvantages of road building in the Amazon?

Explain

3. Why would subsistence farming in the Amazon be referred to as slash-and-burn farming?
4. In what ways would the *environmental changes* of small-*scale* subsistence farming differ from those of large-*scale* soya bean cropping?

Discover

5. Research soya bean farming in the Amazon. How does it compare with cattle ranching in terms of *environmental sustainability*?

Predict

6. Examine the illustration of rainforest destruction shown in figure 3. Draw a sketch of what you predict the area will look like in ten years' time. Use labels and arrows to show important features.

Think

7. Opening up the rainforest with roads can lead to fragmentation of the forest. What might the effect of this be on:
   (a) native animals
   (b) local indigenous populations?
8. Compare how a small-scale farmer from the Amazon and an environmentalist from another country might view the resources of a rainforest.

 **myWorldAtlas** Deepen your understanding of this topic with related case studies and questions.
❷ Forest environments

4.4 SkillBuilder: GIS — deconstructing a map

WHAT IS GIS?

A geographical information system (GIS) is a storage system for information or data, which is stored as numbers, words or pictures.

Go online to access:

- a clear step-by-step explanation to help you master the skill
- a model of what you are aiming for
- a checklist of key aspects of the skill
- a series of questions to help you apply the skill and to check your understanding.

FIGURE 1 GIS allows information to be displayed in a succession of map layers.

Point features marked on first layer of tracing paper.

Line features of rivers and creeks are traced onto a second layer.

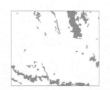

The polygon data of forests are traced onto a third piece of tracing paper.

The three layers of tracing paper are now combined, and BOLTSS are added.

4.5 Paper profits, global losses?

Access this subtopic at **www.jacplus.com.au**

4.6 Should we farm fish?

4.6.1 Why are we overfishing?

The ocean biome has always been seen as an unlimited resource of food for humans. In fact, overfishing is causing the collapse of many of our most important marine ecosystems, and threatens the main source of protein for over one billion people worldwide. **Aquaculture** is a possible solution but, at the same time, it contributes to the decline in fish stocks.

Overfishing is simply catching fish at a rate higher than the fish species can replace themselves. It is an unsustainable use of our oceans and freshwater biomes.

Massive improvements in technology have enabled fish to be located and caught in larger numbers and from deeper, more inaccessible waters. The use of spotter planes, radar and factory ships ensures that fish can be caught, processed and frozen while still at sea.

FIGURE 1 Global protein demand, 1980–2030 (million tonnes)

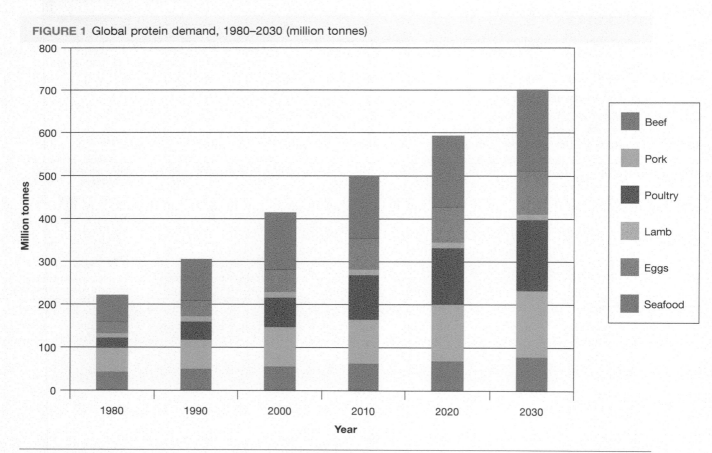

FIGURE 2 Unsustainable fishing

Globally, fish is the most important animal protein consumed (see figure 1). Historically, a lack of conservation and management of fisheries, combined with rising demand for fish products, has seen a 'boom and bust' mentality (see figure 2). The larger fish species are targeted and exploited and, after their populations are decimated, the next species are fished. Examples of these include blue whales, Atlantic cod and bluefin tuna.

4.6.2 What happens when we overfish?

- With overfishing there are often large quantities of by-catch. This means that juvenile fish and other animals, such as dolphins and sea birds, are swept up in nets or baited on hooks before being killed and discarded. For every kilogram of shrimp caught in the wild, 5 kilograms of by-catch is wasted (see figure 3).

- Destructive fishing practices such as cyanide poisoning, dynamiting of coral reefs and bottom trawling (which literally scrapes the ocean floor) cause continual destruction to local ecosystems.

- A large quantity of fish, which could have been consumed by people, is converted to fishmeal to feed the aquaculture industry, and to fatten up pigs, chickens and pet cats (see figure 4).

- Coastal habitats are under pressure. Coral reefs, mangrove wetlands and seagrass meadows, all critical habitats for fish breeding, are being reduced through coastal development, overfishing and pollution.

FIGURE 3 Up to 80 per cent of some fish catches is by-catch.

4.6.3 Shark attack!

Many species of sharks are now threatened with extinction due to excessive overfishing. Their fins are often used to make shark fin soup, an expensive Chinese delicacy and status symbol, and a sign of wealth and prosperity. When sharks are caught, their fins are removed. The live sharks are then thrown overboard and they drown because they cannot swim without fins. This process is known as 'finning'. The United Nations Environment Programme (UNEP) estimates that up to 73 million sharks are killed each year to supply the shark fin market. One consequence of this is that nearly one-third of all ocean-going sharks are now on the internationally recognised list of threatened species. Some populations of hammerhead sharks have declined by 99 per cent in heavily fished regions such as the Mediterranean Sea and the north-west Atlantic Ocean.

FIGURE 4 In Australia, the average cat eats 13.7 kilograms of fish a year compared with the average Australian, who eats 1 kilogram a year.

Sharks are prone to overfishing because they tend to grow and mature slowly and produce relatively few young. They are also often accidently caught up in tuna fishing nets.

The practice of finning is banned in many countries, such as the United States, the European Union and Australia. Instead, the sharks must be brought back to shore before the fins are removed and the bodies must be discarded. This limits the size of catches due to the amount of space available on the boat as well as the need for refrigeration. However, the value of the fins is much higher than the shark meat. For example, an Australian fisherman might only earn $17 per kilogram for flake (shark meat that is filleted and frozen) compared to $880 per kilogram for shark fins, which are simply dried and stacked without refrigeration.

Changing trends

Hong Kong and China have always been the major importers of shark fin, as seen in figure 5, with supplies coming from more than 70 countries across all continents. However, in the past decade the demand and price for fins has decreased by approximately 25 per cent, particularly in China. This decline has been attributed to two main changes. First, in 2013 the Chinese Government banned shark fin soup and other unsustainable seafood from all official events; and second, there has been a major push from environmental organisations such as WildAid to educate people and discourage the consumption of shark fin.

A survey discovered that 75 per cent of Chinese were unaware that the soup, known as 'fish wing soup' was actually made of shark fin, and more than 19 per cent of people interviewed believed that the fins would grow back after being removed. The success of the public awareness campaign has led to more people rejecting finning and shark fin soup.

FIGURE 5 Imports of shark fins into Hong Kong in 2012

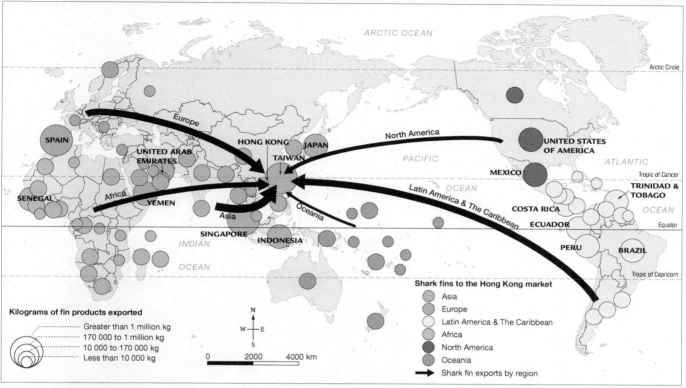

Source: Spatial Vision

What is being done to protect sharks?

- More than 30 airlines are refusing to carry fins as cargo and 15 international hotel chains have dropped the soup from their menus.
- While finning is banned in Australia, it is legal to import, sell and export the fins. Eighteen thousand kilograms were imported in 2015, mostly from New Zealand, Hong Kong, China and the Philippines.
- Shark fishing is legal in India, but the export of fins is banned. Fishermen in coastal India are reliant on shark meat as a low cost protein source. The export of fins brings in much needed money, and fishermen argue that since shark populations have not changed in recent years, the export ban should be lifted.
- Palau, a small country of 22000 people located 800 kilometres east of the Philippines has become the first nation to ban the fishing of sharks and create a shark sanctuary in its surrounding waters. Economically, more can be gained from the tourism industry, particularly scuba diving, than harvesting sharks for meat and fins.
- Mexico, Honduras, the Maldives and other Pacific countries are also in the process of establishing or have already established protected areas for sharks.

4.6.4 Is fish farming the solution?

Aquaculture is one of the fastest growing food industries, providing fish for domestic and export markets. It brings economic benefits and increased food security (see figure 6).

Over 50 per cent of fish consumed by people comes from aquaculture. This is predicted to rise to 60 per cent by 2030. While aquaculture is often seen as a sustainable and eco-friendly solution to overfishing, its rapid growth and poor management in many places has created large-scale environmental change. Some of these changes are described on the next page.

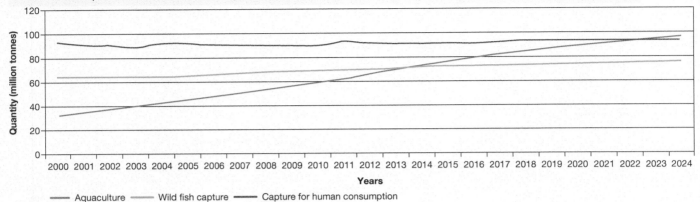

FIGURE 6 Aquaculture and wild fish capture, 2000–2024

Quantity (million tonnes)

Years

— Aquaculture — Wild fish capture — Capture for human consumption

- *Pollution.* Many fish species are fed a diet of artificial food in dry pellets (see figure 7). Chemicals in the feed, and the massive waste generated by fish farms, can pollute the surrounding waters.

- *Loss of fish stock.* Food pellets are usually made of fish meal and oils. Much of this comes from by-catch, but the issue is still that we are catching fish to feed fish. It can take 2 to 5 kilograms of wild fish to produce one kilogram of farmed salmon. Other ingredients in the food pellets include soybeans and peanut meal — products that are suitable for human consumption and grown on valuable farmland.

FIGURE 7 Feeding fish in pens, Thailand

- *Loss of biodiversity.* Many of the fish species farmed are selectively bred to improve growth rates. If accidentally released into the wild, they can breed with native species and so change their genetic makeup. This can lead to a loss of biodiversity. Capture of small ocean fish, such as anchovies, depletes food for wild fish and creates an imbalance in the food chain.

- *Loss of wetlands.* Possibly the greatest impact of aquaculture is in the loss of valuable coastal wetlands. In Asia, over 400 000 hectares of mangroves have been converted into shrimp farms. Coastal wetlands provide important ecological functions, such as protecting the shoreline from erosion and providing breeding grounds for native fish.

4.6 Activities

To answer questions online and to receive **immediate feedback** and **sample responses** for every question, go to your learnON title at www.jacplus.com.au. *Note:* Question numbers may vary slightly.

Remember

1. Refer to figure 2. How important is fish as a source of protein compared with other sources? Use figures in your answer.
2. (a) What is by-catch?
   (b) Examine closely the photograph in figure 3, and describe the by-catch that you see.

3. Refer to figure 6. With the use of figures, compare the predicted growth of wild capture and aquaculture production to 2024.
4. Suggest one reason why wild capture will not increase greatly in the future.
5. Refer to figure 5. Who are the three major exporters of shark fins to Hong Kong? Use figures in your answer.

Explain

6. Explain how overfishing can lead to a loss of biodiversity.
7. (a) What other food items are you aware of that are regarded as status symbols in different cultures? (*Hint:* Consider luxury items.)
   (b) How do you think food items become status symbols in a culture?

Discover

8. Investigate and write a newspaper article on the collapse of the Atlantic cod in Newfoundland. What lessons in the *sustainability* of fishing can be learned from the case of the Atlantic cod?
9. Collect photographs and other information to create an annotated poster showing one of the destructive fishing practices mentioned above.

Predict

10. What do you think the future of aquaculture might be?

 RESOURCES — ONLINE ONLY

 Try out this interactivity: Hook, line and sinker (int-3324)

4.7 SkillBuilder: Interpreting a geographical cartoon

WHAT ARE GEOGRAPHICAL CARTOONS?

Geographical cartoons are humorous or satirical drawings on topical geographical issues, social trends and events. A cartoon conveys the artist's perspective on a topic, generally simplifying the issue.

FIGURE 1 Cartoon on overfishing

Go online to access:

- a clear step-by-step explanation to help you master the skill
- a model of what you are aiming for
- a checklist of key aspects of the skill
- a series of questions to help you apply the skill and to check your understanding.

---

learn on RESOURCES — ONLINE ONLY

Watch this eLesson: Interpreting a geographical cartoon (eles-1731)

 Try out this interactivity: Interpreting a geographical cartoon (int-3349)

4.8 How do we lose land?

4.8.1 What is land degradation?

Land is one of our most basic resources and one that is often overlooked. In our quest to produce as much as possible from the same area of land, we have often failed to manage it sustainably. Land degradation is the result of such poor management.

Land degradation is a decline in the quality of the land to the point where it is no longer productive. Land degradation covers such things as soil **erosion**, invasive plants and animals, **salinity** and desertification. Degraded land is less able to produce crops, feed animals or renew native vegetation. There is also a loss in soil fertility because the top layers, rich in **humus**, can be easily eroded by wind or water. In Australia, it can take up to 1000 years to produce just three centimetres of soil, which can be lost in minutes in a dust storm.

Globally, degradation has caused the loss of more than 350 million square kilometres of the Earth's surface. In Australia, of the five million square kilometres of land used for agriculture, more than half has been affected by, or is in danger of, degradation.

4.8.2 What are the causes of land degradation?

Land degradation is common to both the developed and developing world, and results from both human and natural causes.

Human causes

Human causes of land degradation involve unsustainable land management practices, such as:
- land clearance — deforestation or excessive clearing of protective vegetation cover
- overgrazing of animals — plants are eaten down or totally removed, exposing bare soil, and hard-hoofed animals such as cows and sheep compact the soil (see figure 1)
- excessive irrigation — can cause watertables to rise, bringing naturally occurring salts to the surface, which pollute the soil
- introduction of exotic species — animals such as rabbits and plants such as blackberries become the dominant species
- decline in soil fertility — caused by continual planting of a single crop over a large area, a practice known as monoculture
- farming on marginal land — takes place on areas such as steep slopes, which are unsuited to ordinary farming methods.

FIGURE 1 Soil erosion as a result of overgrazing in Australia

Biophysical causes

Natural processes such as prolonged drought can also lead to land degradation. However, land can sometimes recover after a drought period. Topography and the degree of slope can also influence soil erosion. A steep slope is more prone to erosion than flat land.

4.8.3 What are the impacts of land degradation?

As land becomes degraded, productivity, or the amount of food it can produce, is lost. Some countries in Sub-Saharan Africa have lost up to 40 per cent productivity in croplands over two decades, while population has doubled in the same time period. Farmers may choose to abandon the land, try to restore the land or, if the pressure to produce food is too great, they may have no choice but to continue using the land. Unproductive land will be exposed to continual erosion or weed invasion.

If extra fertilisers are applied to try to improve fertility, the excessive nutrients can create pollution and algae build-up in nearby streams. Airborne dust creates further hazards for both people and air travel. Land degradation is a classic example of human impact on all spheres of the environment — atmosphere, biosphere, lithosphere and hydrosphere.

Figure 2 shows the total amount of land across the Earth's surface (not just farmland) and the extent of land degradation. About 40 per cent of degraded lands are found in places that experience widespread poverty, which is a contributing factor to food insecurity. Poor farmers with degraded land and few resources often have little choice but to continue to work the land.

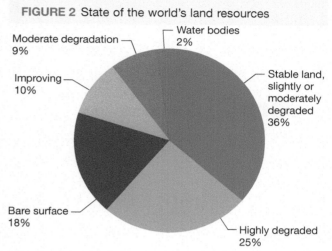

FIGURE 2 State of the world's land resources

- Moderate degradation 9%
- Improving 10%
- Bare surface 18%
- Water bodies 2%
- Stable land, slightly or moderately degraded 36%
- Highly degraded 25%

Source: Food and Agriculture Organization of the United Nations 2011

4.8.4 Are we turning drylands into deserts?

Desertification is an extreme form of land degradation. It usually occurs in semi-arid regions of the world, and the result gives the appearance of spreading deserts. Desert biomes, or arid regions, are harsh, dry environments where few people live. In contrast, semi-arid regions, or drylands, occupy 41 per cent of the Earth's surface and support over two billion people, 90 per cent of whom live in developing nations. While traditional grazing and cropping has taken place in dryland regions for centuries, population growth and the demand for food has put enormous pressure on land resources. Overclearing of vegetation, overgrazing and overcultivation are a recipe for desertification.

4.8 Activities

To answer questions online and to receive **immediate feedback** and **sample responses** for every question, go to your learnON title at www.jacplus.com.au. *Note:* Question numbers may vary slightly.

Remember

1. List two human and two natural causes of land degradation.
2. Refer to figure 2. What percentage of the world's land resources are classified as moderately to highly degraded?
3. Which biome supports more life: desert or drylands? Why?

Explain

4. Create an annotated sketch to show the *interconnection* between plants and soil. Use the following points as labels on your sketch.
   - Plant roots help hold soil together.
   - Decomposing plants add nutrients to the soil.
   - Plants shade the topsoil and reduce evaporation.
   - Plants reduce the speed of wind passing over the ground.

5. Explain how land degraded by drought may recover, whereas land degraded by cultivation may not.
6. Study the photograph in figure 1. Why would it be difficult to either graze animals or grow crops on this land?

Discover

7. Investigate an area in Victoria that is suffering from land degradation. Identify the location, causes and impacts of the degradation. Are any steps being taken to reduce the impacts?

Think

8. Examine the photograph in figure 1 again. If this was your property and your livelihood, what steps would you take to reduce the erosion problem?

learn **on** RESOURCES — ONLINE ONLY

✦ **Try out this interactivity:** Losing land (int-3325)

4.9 Irrigation: success or failure?

4.9.1 What is the purpose of irrigation?

Food production and security is directly related to water availability. As population increases, so too does demand for water. Moreover, there are always competing demands for water from the domestic, industrial and environmental sectors. In many places in the world, water is becoming increasingly scarce. Consequently, the development of water resources is becoming more expensive and, in some cases, environmentally destructive.

Most of the world's food production is rain-fed; that is, dependent on naturally occurring rainfall. Only a small proportion of agricultural land is irrigated, yet **irrigation** is now the biggest user of water in the world, consuming 70 per cent of the world's freshwater resources. Irrigation brings many benefits, such as:

- supplementing or replacing rain, especially in places where rainfall is low or unreliable. In many parts of the world, it is not possible to produce food without irrigation.
- increasing crop yields, up to three times higher than rain-fed crops. Only 20 per cent of the world's farmland is irrigated but it produces over 40 per cent of our food.
- enabling a wide variety of foods to be grown, especially those with high water needs, such as rice, or with high value, such as fruit and wine grapes
- flexibility, being used at different times according to crop needs; for example, during planting and growing or close to harvest time.

FIGURE 1 Irrigation allows for pasture to be grown in times of drought. Compare the irrigated with the non-irrigated paddocks.

4.9.2 What are the impacts of irrigation on the environment?

The benefits of irrigation have resulted in increased food production and greater food security. But irrigation has created major changes to the biomes where it is used. Irrigation changes the natural environment by extracting water from rivers and lakes and building structures to store, transfer and dispose of water. The topography or shape of the land is often changed too, as occurs when terraces are built for paddy fields. In addition, irrigation water is often applied to the land in much larger quantities than naturally occurs, which can then change soil composition and cause **waterlogging** and salinity problems.

How does irrigation create salinity problems?

On irrigated land, salinity is the major cause of land being lost to production, which is an area the size of France or 62 million hectares of lost productive land worldwide. It is also a major cause of land degradation in Australia (see figure 2).

Overwatering of shallow-rooted crops adds excess water to the **watertable**, causing it to rise (see figure 3). If the subsoils are naturally salty, much of this salt can be drawn to the surface. Most crops and pasture will not grow in salty soils, so the land becomes useless for farming. Land that is affected by salinity is also more prone to wind and water erosion.

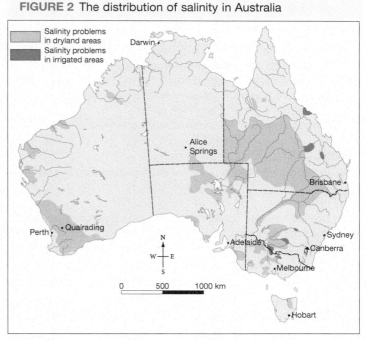

FIGURE 2 The distribution of salinity in Australia

Salinity problems in dryland areas
Salinity problems in irrigated areas

Source: Spatial Vision.

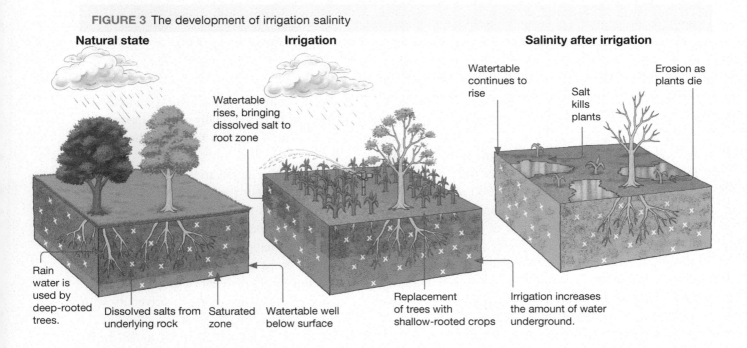

FIGURE 3 The development of irrigation salinity

Natural state

Rain water is used by deep-rooted trees.

Dissolved salts from underlying rock

Saturated zone

Irrigation

Watertable rises, bringing dissolved salt to root zone

Watertable well below surface

Replacement of trees with shallow-rooted crops

Irrigation increases the amount of water underground.

Salinity after irrigation

Watertable continues to rise

Salt kills plants

Erosion as plants die

4.9 Activities

To answer questions online and to receive **immediate feedback** and **sample responses** for every question, go to your learnON title at www.jacplus.com.au. *Note*: Question numbers may vary slightly.

Remember

1. What is meant by the term *waterlogging*?
2. What percentage of the world's fresh water is consumed by irrigation? What would be the other main uses of water?

Explain

3. (a) Study the map in figure 2, showing the distribution of salinity in Australia. Estimate the approximate percentage of each state affected by salinity.
   (b) Why do you think dryland salinity covers a larger area than irrigation salinity?
4. Use figure 3 as a model and create a similar sketch. Annotate your drawing with suggestions on how to reduce the effects of irrigation salinity.
5. What **changes** to the **environment** are needed in order to irrigate a large region?

Discover

6. Investigate methods used in Australia to reduce the **environmental** effects of salinity.

Think

7. Soil salinity was not a problem when Indigenous Australian peoples were the land's sole caretakers. What does this suggest about land management practices in this country since 1788?
8. Has irrigation been a success or failure? Write a paragraph expressing your viewpoint.

4.10 Does farming use too much water?

Access this subtopic at **www.jacplus.com.au**

4.11 Why is global biodiversity diminishing?

4.11.1 The loss of biodiversity

The last few centuries have seen the greatest rate of species extinction in the history of the planet (see figure 1). The population of most species is decreasing, and genetic diversity is declining, especially among species that are cultivated for human use. Six of the world's most important land biomes have now had more than 50 per cent of their area converted to agriculture (see figure 2).

In those places where there has been very little industrial-scale farming, a huge variety of crops are still grown. In Peru, for example, over 3000 different potatoes are still cultivated. Elsewhere, biodiversity as well as agricultural biodiversity (biodiversity that is specifically related to food items) is in decline. In Europe, 50 per cent of all breeds of domestic animals have become extinct, and in the United States, 6000 of the original 7000 varieties of apple no longer exist. How has this happened?

- Industrial-scale farming and new high-yielding, genetically uniform crops replace thousands of different traditional species. Two new rice varieties in the Philippines account for 98 per cent of cropland.
- Converting natural habitats to cropland and other uses replaces systems that are rich in biodiversity with monoculture systems that are poor in diversity (see figure 3).
- Industrial-scale farming and new high-yielding, genetically uniform crops replace thousands of different traditional species. Two new rice varieties in the Philippines account for 98 per cent of cropland.

FIGURE 1 Extinctions per thousand species per millennium

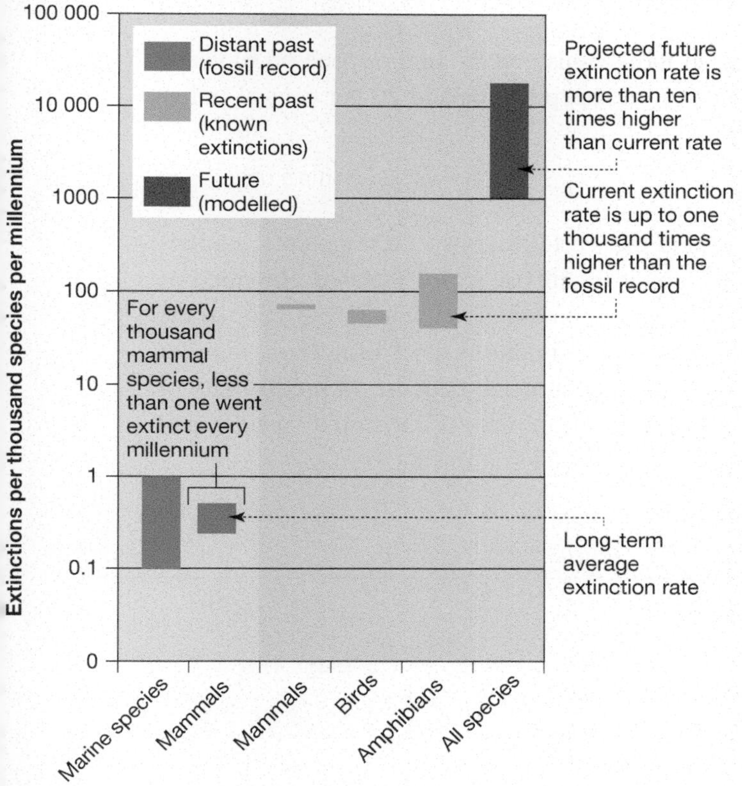
100 000
10 000
1000
100
10
1
0.1
0
Extinctions per thousand species per millennium
Distant past (fossil record)
Recent past (known extinctions)
Future (modelled)
Projected future extinction rate is more than ten times higher than current rate
Current extinction rate is up to one thousand times higher than the fossil record
For every thousand mammal species, less than one went extinct every millennium
Long-term average extinction rate
Marine species
Mammals
Mammals
Birds
Amphibians
All species

FIGURE 2 Percentage of biomes converted to agriculture over time

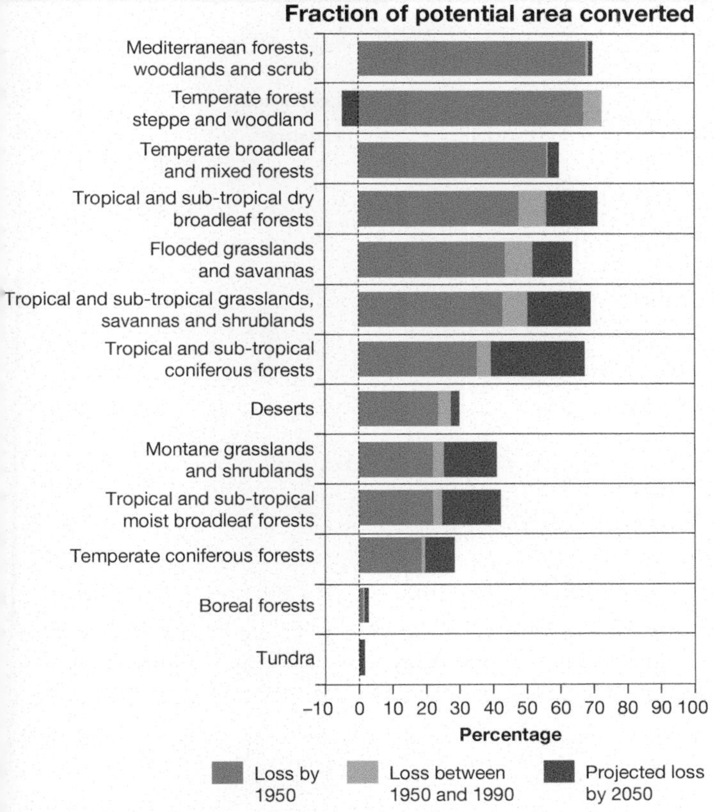
Fraction of potential area converted
Mediterranean forests, woodlands and scrub
Temperate forest steppe and woodland
Temperate broadleaf and mixed forests
Tropical and sub-tropical dry broadleaf forests
Flooded grasslands and savannas
Tropical and sub-tropical grasslands, savannas and shrublands
Tropical and sub-tropical coniferous forests
Deserts
Montane grasslands and shrublands
Tropical and sub-tropical moist broadleaf forests
Temperate coniferous forests
Boreal forests
Tundra
−10 0 10 20 30 40 50 60 70 80 90 100
Percentage
Loss by 1950
Loss between 1950 and 1990
Projected loss by 2050

FIGURE 3 Changes to percentage of original species according to changes in biomes for food production

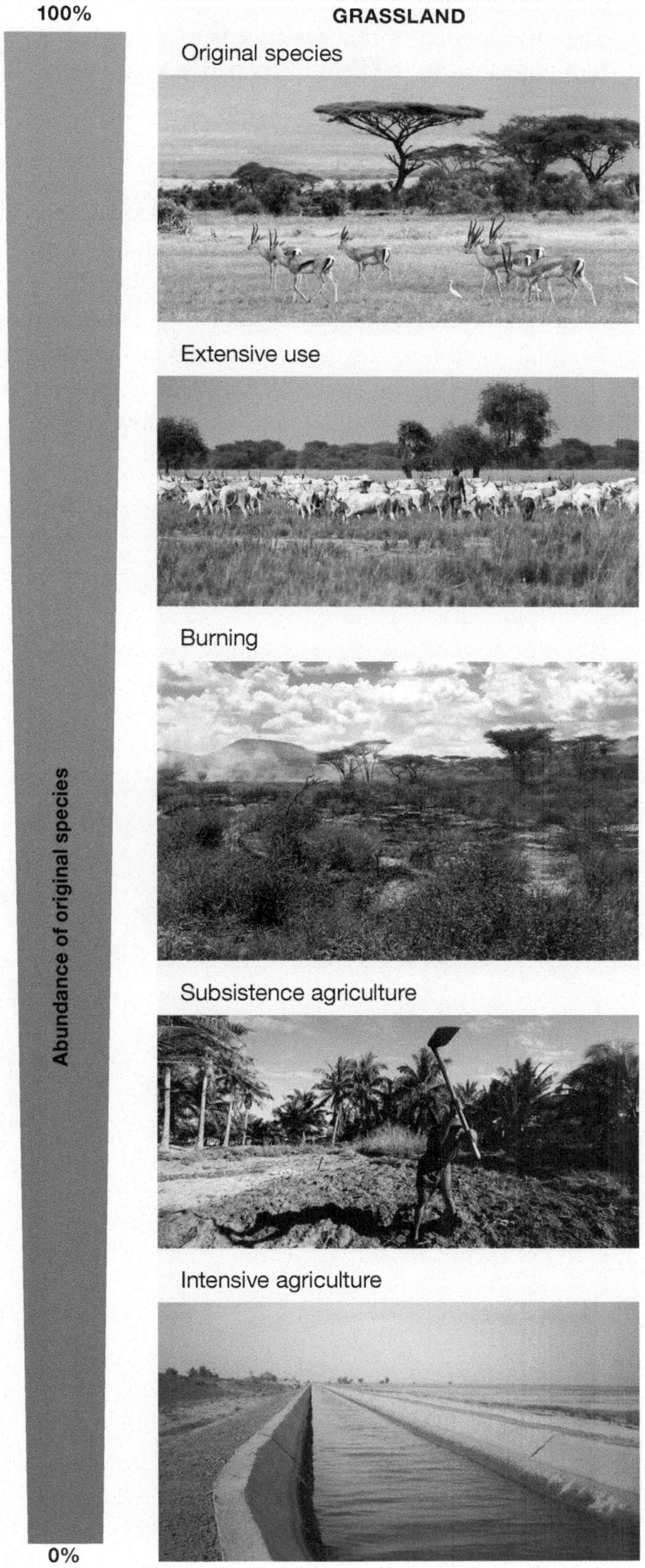
100%
GRASSLAND
Original species
Extensive use
Burning
Subsistence agriculture
Intensive agriculture
Abundance of original species
0%

- Uniform crops are vulnerable to pests and diseases, which then require large inputs of chemicals that ultimately pollute the soil and water. Traditional ecosystems have many natural enemies to combat pest species.
- The introduction of modern breeds of animals has displaced indigenous breeds. In the space of 30 years, India has lost 50 per cent of its native goat breeds, 30 per cent of sheep breeds and 20 per cent of indigenous cattle breeds.

4.11.2 Australia's biodiversity

Australia has a high number of **endemic** species, and 7 per cent of the world's total species of plants, animals and micro-organisms. That makes Australia one of only 17 countries in the world that are classified as megadiverse — having high levels of biodiversity. These 17 nations combined contain 75 per cent of the Earth's total biodiversity (see figure 4). Australia's unique biodiversity is due to its 140 million years of geographic isolation. However, Australia has experienced the largest documented decline in biodiversity of any continent over the past 200 years. It is thought that 50 species of animals (27 mammal species and 23 bird species) and 48 plant species are now extinct.

FIGURE 4 Distribution of megadiverse countries

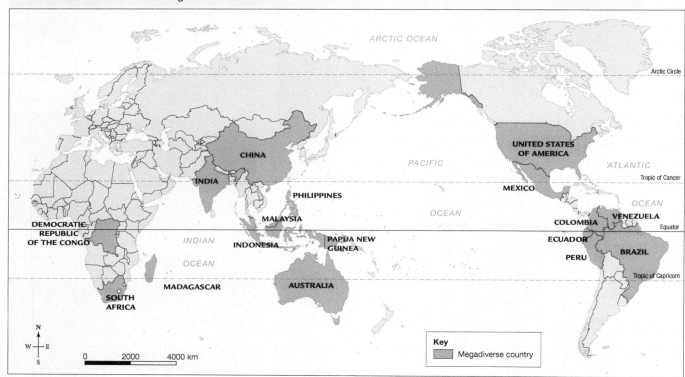

Source: Spatial Vision

4.11 Activities

To answer questions online and to receive **immediate feedback** and **sample responses** for every question, go to your learnON title at www.jacplus.com.au. *Note*: Question numbers may vary slightly.

Remember

1. Describe the ways in which human activities can lead to a loss in biodiversity.
2. What is a megadiverse country? Why is Australia considered a megadiverse country?

Explain

3. (a) Study figure 2. Which three biomes have seen the greatest percentage *change* in areas converted to cultivation? Use figures in your answer.
   (b) Suggest why these three have had the most *change*.
4. Study the information in figure 3. Describe the *changes* to the grassland biome as seen over time.

Think

5. In what ways would the Indigenous Australian peoples' practice of rotational land occupation have helped maintain biodiversity before European occupation?
6. Does it matter that we have fewer species of apples or goats?

4.12 Does farming cause global warming?

Access this subtopic at **www.jacplus.com.au**

4.13 Review

4.13.1 Review

The Review section contains a range of different questions and activities to help you revise and recall what you have learned, especially prior to a topic test.

4.13.2 Reflect

The Reflect section provides you with an opportunity to apply and extend your learning.

Access this subtopic at **www.jacplus.com.au**

TOPIC 5
Are we devouring our future?

5.1 Overview

Numerous **videos** and **interactivities** are embedded just where you need them, at the point of learning, in your learnON title at www.jacplus.com.au. They will help you to learn the content and concepts covered in this topic.

5.1.1 Introduction

There is enough food produced in the world today to feed every man, woman and child, so why is it that one in nine people, or 795 million, will go to bed hungry tonight? What is stopping everyone getting enough to eat? If this is the current situation, what is going to happen in the future?

For these children, in a tent camp for people displaced by flooding in northern India, the only kind of food security is in the form of aid.

Starter questions

1. How long has it been since you had anything to eat?
2. How many different food items have you eaten today?
3. How many of these did your family grow?
4. Do you know when and where your next meal is coming from?
5. Do you feel secure in knowing that you have food in your home?
6. How would your answers to these questions compare with those of the people shown in the photograph on the previous page?
7. Why do you think we have so many people hungry when there is enough food produced in the world?
8. How will the world feed its future population?

INQUIRY SEQUENCE

5.1	Overview	88
5.2	Who's not hungry?	89
5.3	Who is hungry?	92
5.4	**SkillBuilder:** Constructing and describing complex choropleth maps	online only · 95
5.5	How does a famine develop?	online only · 95
5.6	Will land loss lead to food shortages?	95
5.7	**SkillBuilder:** Interpreting satellite images to show change over time	online only · 100
5.8	Are we running dry?	100
5.9	Climate change: freeze or fry?	103
5.10	Why is food being wasted?	online only · 106
5.11	**Review**	online only · 106

5.2 Who's not hungry?

5.2.1 What is food security?

Very few Australians, by choice, would go to bed at night hungry. We live in a country where there is a plentiful supply and wide range of food items available. Our relatively high standard of living enables most of us to afford to purchase, store and prepare food, or even dine out. Most of us are secure in the knowledge that there will be food at the next meal time.

Food security exists when all people, at all times, have physical and economic access to enough safe and nutritious food to meet their dietary needs and food preferences for an active and healthy lifestyle.

Source: Food and Agriculture Organization

Food security for you, as a student, means that your family either grows its own food, has sufficient income to purchase food, or is able to barter or swap food. Similarly, food security for a country means that it is able to grow sufficient food, or it has enough wealth to import food, or it combines the two. Not all people in the world are able to achieve this. For example, consider the range of foods available in the two markets in figures 1 and 2.

FIGURE 1 The fresh produce section of a supermarket in a developed economy

FIGURE 2 A food market in a developing economy

5.2.2 Who has food security?

Figure 3 shows the countries of the world regarded as having low risk of food insecurity shaded in dark green. This is based on a range of 12 different **indicators**, including the:

• affordability of food
• accessibility of food
• nutritional value of food
• safety of food
• nutritional and health status of the population.

Countries that have low risk of food security are able to produce more food than they require, so they can export their surplus; or they are able to afford to import all their needs, as is the case for the United Arab Emirates, for example.

Australia, for instance, produces three times as much food as it consumes, and is a major exporter of both fresh and processed food. We can trade competitively in cereals, oil seeds, beef, lamb, sugar and dairy products. Ninety per cent of our food is grown in Australia. Of the remaining 10 per cent that we import, many foods are either processed or out of season; oranges are an example. Global trade is an important component of food security because it is almost impossible to exactly match food production to food demands.

As a country, Australia does not have a problem feeding its population but it has a humanitarian interest in the food security of developing nations. As a major food producer, Australia does face future challenges. There is declining growth in agricultural productivity, the threat of climate change, and increasing competition for land and water.

FIGURE 3 The Food Security Risk Index, 2013

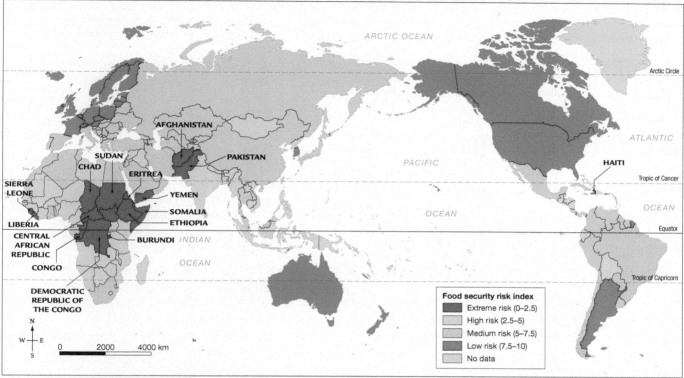

Source: Verisk Maplecroft — Verisk Maplecroft's Food Security Index provides a quantitative assessment of risks to the continued availability, stability and access to sufficient food supplies. The index also considers the nutritional outcomes of each country's relative food security.

5.2 Activities

To answer questions online and to receive **immediate feedback** and **sample responses** for every question, go to your learnON title at www.jacplus.com.au. *Note*: Question numbers may vary slightly.

Remember

1. What is meant by the term *food security*?
2. How do you give your pet food security?
3. Refer to figure 3 and your atlas.
    (a) List five examples of countries, from different regions of the world, that are considered to have low risk of food insecurity.
    (b) Would you classify these countries as developing or developed?
    (c) What does it mean for people who live in a country with low risk of food insecurity?

Explain

4. Compare the two photographs in figures 1 and 2.
    (a) What are the similarities and differences between the two markets? Do you think all food groups would be available in both markets? Why or why not?
    (b) List 10 food items (not including water) that you could live on for a week, while still maintaining a balanced diet. You are not allowed anything else.
5. The Food Security Risk Index was based on evaluating five different indicators. Why do you think indicators such as accessibility and safety were included?

Discover

6. (a) Use the **Australian food statistics** weblink in the Resources tab to construct a graph of the 10 largest food imports and exports in terms of dollar value.
   (b) Describe your completed graph, using figures in your answer.

Predict

7. How do you think climate *change* might affect Australia's food security?
8. What natural or human events could disrupt our food security?

learnon RESOURCES — ONLINE ONLY

🔗 **Explore more with this weblink:** Australian food statistics

5.3 Who is hungry?

5.3.1 Who is at risk of food insecurity?

Have you ever felt hungry? How long since you last had a meal? How long before your next meal? How easy is it to open the fridge or cupboard for a snack? But what if there was no food in the house, no money to purchase food with and very little food available in the shops? If that was the case, you would be just one of the hundreds of millions of other people in the world in a similar position.

Figure 1 highlights those countries that are most at risk of not having reliable access to food. Seventy five per cent of all hungry people live in rural areas, mainly in rural environments of Asia and Africa. These people are often totally dependent on agriculture for food, and have no alternative source of work or income. Consequently they are vulnerable to food insecurity.

FIGURE 1 The Food Security Risk Index, 2013

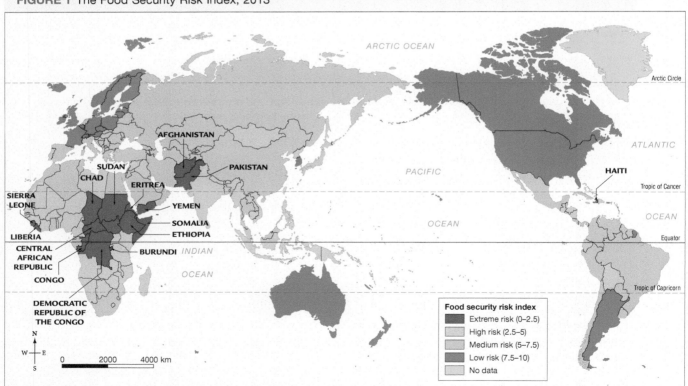

Source: Verisk Maplecroft — Verisk Maplecroft's Food Security Index provides a quantitative assessment of risks to the continued availability, stability and access to sufficient food supplies. The index also considers the nutritional outcomes of each country's relative food security.

5.3.2 What happens when people do not have food security?

For the 800 million people who do not have enough to eat, the issue of finding sufficient and nutritious food must be faced daily. At least 2 billion of the world's people are **undernourished**, with diets that are minimal or below the level of sustenance. People who do not have a regular and healthy diet often have shortened life expectancy and an increased risk of disease. Children are especially vulnerable to poor diet, and their growth, weight, and physical and mental development suffer. In India, 29.4 per cent of children are underweight while 38.4 per cent have stunted growth, indicating they are **malnourished**. In the previous decade this figure was closer to 48 per cent of the child population.

5.3.3 Why is there food insecurity?

Global food production now provides 17 per cent more calories per head than 30 years ago, despite the world's population increasing by 70 per cent. There is, however, unequal access to **arable** land, technology, education and employment opportunities. In the past decade the actual number of undernourished people in the world has fallen by 216 million, mostly due to large-scale improvements in the highly populated countries of India and China. Improvements in food production and economic development have not always occurred in those places experiencing rapid growth in population. Food is redistributed around the world via trade and aid but neither is a long-term or large-scale solution to food insecurity. Regional variations still occur in the distribution of hunger as can be seen in figure 3.

Some of the reasons for food insecurity include:
- poverty
- population growth
- weak economy and/or political systems
- conflict
- natural disasters such as drought.

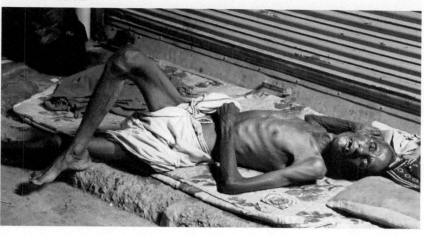

FIGURE 2 An old man in Delhi suffering from a lifetime of malnutrition

FIGURE 3 The changing distribution of hunger in the world: numbers and shares of undernourished people by region, 1990–92 and 2014–16

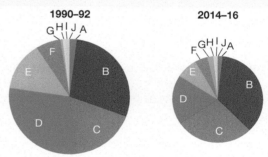

1990–92
Total = 1010 million

2014–16
Total = 795 million

		Number (millions)		Regional share (%)	
		1990–92	2014–16	1990–92	2014–16
A	Developed regions	20	15	2.0	1.8
B	Southern Asia	291	281	28.8	35.4
C	Sub-Saharan Africa	176	220	17.4	27.7
D	Eastern Asia	295	145	29.2	18.3
E	South-Eastern Asia	138	61	13.6	7.6
F	Latin America and the Caribbean	66	34	6.5	4.3
G	Western Asia	8	19	0.8	2.4
H	Northern Africa	6	4	0.6	0.5
I	Caucasus and Central Asia	10	6	0.9	0.7
J	Oceania	1	1	0.1	0.2
	Total	**1 011**	**795**	**100**	**100**

Note: The areas of the pie charts are proportional to the total number of undernourished in each period. Data for 2014–16 refer to provisional estimates. All figures are rounded.

Source: FAO

5.3.3 Activities

To answer questions online and to receive **immediate feedback** and **sample responses** for every question, go to your learnON title at www.jacplus.com.au. *Note*: Question numbers may vary slightly.

Remember

1. What factors make people vulnerable to food insecurity?
2. What is the difference between undernutrition and malnutrition?
3. Use the **Malnutrition** weblink in the Resources tab and watch the video. What is the link between malnutrition in children and the mother's health?

Explain

4. Refer to a map of conflict in your atlas or online. Is there an *interconnection* between those countries that have a high or extreme risk of food insecurity and those countries that are experiencing conflict? Include country names in your answer.
5. Explain how conflict can lead to food insecurity.

Discover

6. Research and find out the causes and effects of one of the conditions caused by dietary deficiency, such as deficiency in iron, vitamin A or vitamin C.
7. Select one of the *places* mapped in figure 1 as being extreme risk. Find out the main factors that contribute to its food insecurity.
8. Refer to figure 3. Compare the three regions of the world that had the highest number of undernourished people in 1990–92 and 2014–16.
    (a) Which region of the world has shown the greatest decrease in the percentage of undernourished people?
    (b) Which region has shown the greatest increase in the percentage of undernourished people?

Predict

9. With a partner, develop five steps you think would reduce a country's risk of food insecurity. Give reasons for your choices.
10. How can Australia best help another country that is at high risk of having insufficient food for its people?

Think

11. At the turn of the twentieth century, the total worldwide spending on agricultural research was US$23 billion. Compare this to $1.5 trillion on weapons. Do we have our priorities right? Write a short letter to the editor outlining your viewpoint.

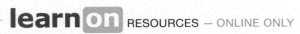 RESOURCES — ONLINE ONLY

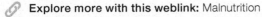

🔗 **Explore more with this weblink:** Malnutrition

🧩 **Try out this interactivity:** Nothing to eat (int-3326)

5.4 SkillBuilder: Constructing and describing complex choropleth maps

WHAT IS A COMPLEX CHOROPLETH MAP?

A complex choropleth map is a map that is shaded or coloured to show the average density or concentration of a particular feature or variable, and it shows an area in detail.

Go online to access:

- a clear step-by-step explanation to help you master the skill
- a model of what you are aiming for
- a checklist of key aspects of the skill
- a series of questions to help you apply the skill and to check your understanding.

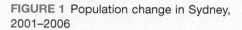

FIGURE 1 Population change in Sydney, 2001–2006

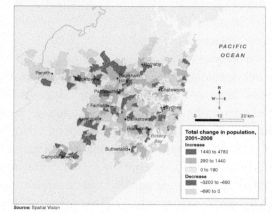

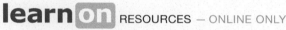

learn on RESOURCES — ONLINE ONLY

📹 **Watch this eLesson:** Constructing and describing complex choropleth maps (eles-1732)

➡️ **Try out this interactivity:** Constructing and describing complex choropleth maps (int-3350)

5.5 How does a famine develop?

Access this subtopic at **www.jacplus.com.au**

5.6 Will land loss lead to food shortages?

5.6.1 How is land lost?

Land is absolutely essential for food production, and the world has more than enough arable land to meet future demands for food. Nevertheless, we need to find a balance between competing demands for this finite resource.

The loss of productive land has two main causes. First, there is the degradation of land quality through such things as erosion, **desertification** and salinity. Second, there is the competition for land from non-food crops, such as biofuels, and from expanding urban areas. As figure 1 shows, the growth in world population is inversely proportional to the amount of arable land available. This does not even take into consideration the land that is degraded and no longer suitable for growing food.

Land degradation

Although there have been significant improvements in crop yields, seeds, fertilisers and irrigation, they have come at a cost. Environmental degradation of water and land resources places future food production at risk.

The main forms of land degradation are:
- erosion by wind and water
- salinity
- pest invasion
- loss of biodiversity
- desertification.

Land degradation occurs in all food-producing biomes across the globe. Some degradation occurs naturally; for example, a heavy rainstorm can easily wash away topsoil. However, the most extensive degradation is caused by overcultivation, overgrazing, overwatering, overloading with chemicals and overclearing (see figure 2). Currently, 25 per cent of the world's land is highly degraded, while only 10 per cent is improving in quality. In South-East Asia, 50 per cent of cultivated land has severe soil quality problems, which prevent increases in food production. The Ministry for Agriculture in China estimates that 3.3 million hectares of arable land is polluted with chemicals and heavy metals, mostly in regions that grow grains.

Competition for land

There has been a growing global trend to convert valuable cropland to other uses. Urban growth, industrialisation and energy production all require land. For example, in less than 16 years, China lost more than 14.5 million hectares of arable land to other land uses. This land no longer produces food, which then puts pressure on existing land resources to make up the loss.

Growing fuel

Traditionally, the main forms of biofuel have been wood and charcoal. Almost 90 per cent of wood harvested in Africa and 40 per cent harvested in Asia is used for heating and cooking. Today, people are seeking more renewable energy sources and wanting to reduce CO_2 emissions associated with deforestation, so there is greater demand for alternative energy sources. Consequently, the use of agricultural crops to produce biofuels is increasing. Ethanol (mostly used as a substitute for petrol) is extracted from

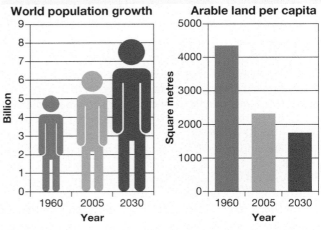

FIGURE 1 Comparison of world population growth and arable land per capita

FIGURE 2 Land degradation caused by deforestation in Madagascar

crops such as corn, sugar cane and cassava. Biodiesel is derived from plantation crops such as palm oil, soya beans and **jatropha**. The growth of the biofuel industry has the potential to threaten future food security by:
- changing food crops to fuel crops, so less food is produced and crops have to be grown on **marginal land** rather than arable land
- increasing prices, which makes staple foods too expensive for people to purchase
- forcing disadvantaged groups, such as women and the landless poor, to compete against the might of the biofuel industry.

Creeping cities

Cities tend to develop in places that are agriculturally productive. However, as they expand, they encroach on valuable farmland.

FIGURE 3a Satellite image of the city of Tehran in 1985

FIGURE 3b Satellite image of Tehran in 2009 — the expansion of the city has taken over valuable arable land.

Land grabs

A growing challenge to world food security is the purchase or lease of land, largely in developing nations, by resource-poor but wealthier nations. Large-scale 'land grabs', as they are known, have the potential to improve production and yields but at the same time there is growing concern over the loss of land rights and food security for local populations. Since 2002, over 36 million hectares, an area bigger than Victoria and Tasmania combined, has been purchased or leased by foreigners around the world, with another 15 million hectares being negotiated. By and large, the land buyers are dominated by four regions: western Europe, North America, the Middle East and the expanding economies within Asia. China has land in 33 different countries, the United States in 28 and the United Kingdom in 30. Australia has land in nine different countries, while 13 countries have control over land in Australia.

The regions where land is purchased tend to be located in Africa, South America, eastern Europe and parts of South East Asia. Ethiopia 'exports' land to 21 different countries, while Madagascar and the Philippines export to 18 countries. Figure 4 shows the global scale of land grabs.

The rise of land grabs came about as a result of the 'triple-F' crisis — food, fuel and finance.
- *Food crisis*: massive increases in world food prices in 2007–08 emphasised the need for those countries heavily reliant on importing food, such as Saudi Arabia and China, to improve their food security by obtaining land in other countries to produce food to meet their own needs.
- *Fuel crisis*: rising and fluctuating oil prices in 2007–09 created an incentive for countries to acquire land to produce their own biofuels (see figure 5).
- *Financial crisis*: the global financial crisis in 2008 saw organisations switch from investing in stocks and shares to land in overseas countries, especially land that could be used to generate food and fuel crops.

FIGURE 4 Land grabs — who's buying where

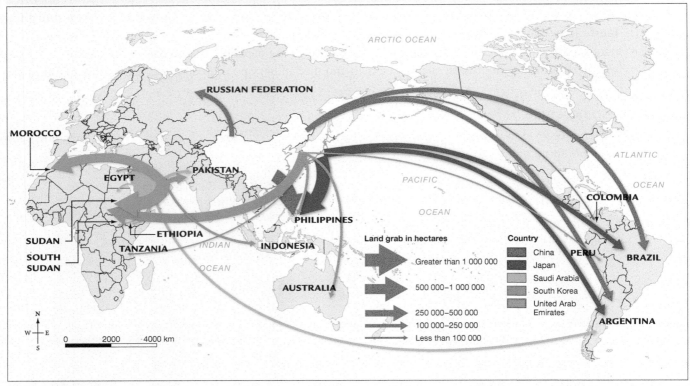

Source: Spatial Vision

FIGURE 5 Countries in Africa where land is being bought by other nations for biofuel crops

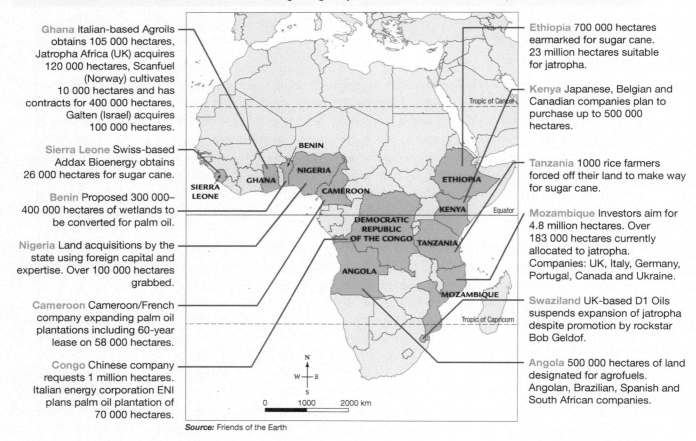

Ghana Italian-based Agroils obtains 105 000 hectares, Jatropha Africa (UK) acquires 120 000 hectares, Scanfuel (Norway) cultivates 10 000 hectares and has contracts for 400 000 hectares, Galten (Israel) acquires 100 000 hectares.

Sierra Leone Swiss-based Addax Bioenergy obtains 26 000 hectares for sugar cane.

Benin Proposed 300 000–400 000 hectares of wetlands to be converted for palm oil.

Nigeria Land acquisitions by the state using foreign capital and expertise. Over 100 000 hectares grabbed.

Cameroon Cameroon/French company expanding palm oil plantations including 60-year lease on 58 000 hectares.

Congo Chinese company requests 1 million hectares. Italian energy corporation ENI plans palm oil plantation of 70 000 hectares.

Ethiopia 700 000 hectares earmarked for sugar cane. 23 million hectares suitable for jatropha.

Kenya Japanese, Belgian and Canadian companies plan to purchase up to 500 000 hectares.

Tanzania 1000 rice farmers forced off their land to make way for sugar cane.

Mozambique Investors aim for 4.8 million hectares. Over 183 000 hectares currently allocated to jatropha. Companies: UK, Italy, Germany, Portugal, Canada and Ukraine.

Swaziland UK-based D1 Oils suspends expansion of jatropha despite promotion by rockstar Bob Geldof.

Angola 500 000 hectares of land designated for agrofuels. Angolan, Brazilian, Spanish and South African companies.

Source: Friends of the Earth

The risk to food security

Investors in farmland are, understandably, seeking large expanses of land that has fertile soils and good rainfall or access to irrigation water. Often, land that is purchased is already occupied and used by small-scale farmers, in particular women who rarely benefit from any compensation. Prices for land can be much lower and there is frequently corruption, with much money going to local and government officials. People can also be forced off their land by governments keen to make deals with wealthy governments and corporations. Many land grabs have neglected the social, economic and environmental impacts of the deals.

With the purchase of land can come the right to withdraw the water linked to it and this can deny local people access to water for fishing, farming and spraying animals. Withdrawal of water can reduce flow downstream. The Niger River, West Africa's largest river, flows through three countries and sustains over 100 million people, so any large-scale water reductions create significant impact to downstream environments and people. One country the Niger River flows through is Mali. By the end of 2010, about 500 000 hectares had been leased by foreign companies and countries in Mali. Although Mali has limited arable land and two million people are underfed, a high percentage of land bought by foreign owners will be used for biofuel crops.

It has been estimated that the land taken up by foreign investors could feed as many as 190 to 370 million people, or even more, if yields are raised to the level of industrialised western faming. As well, there are environmental risks associated with monoculture farming and the loss of biodiversity in the region.

5.6 Activities

To answer questions online and to receive **immediate feedback** and **sample responses** for every question, go to your learnON title at www.jacplus.com.au. *Note:* Question numbers may vary slightly.

Remember
1. What are the two main ways that productive farmland can be lost?
2. Why is the use of corn as a biofuel a threat to food security?
3. What is meant by the term *land grab*?

Explain
4. Refer to figure 1.
   (a) Describe the *changes* in population growth and the arable land per person between 1960 and 2030, making use of figures.
   (b) What do these graphs suggest about food security?
5. Compare the advantages and disadvantages in developing and developed nations of using traditional biofuels, such as wood and charcoal, instead of oil and gas.
6. Refer to figure 4.
   (b) Which three countries are the largest purchasers of overseas land? Use figures in your answer.
   (b) Why do you think South Korea has invested in so many countries in such different *places*?

Discover
7. What is jatropha? What are the benefits of growing this rather than corn and other biofuels?
8. What is happening in Australia? Investigate which foreign companies own farmland here, what they are using it for and where it is located.

Predict
9. Do you think Australia will need to purchase farmland overseas? Give reasons for your answer.

Think
10. Are land grabs an effective solution for establishing a country's food security? Discuss your point of view.

 RESOURCES — ONLINE ONLY

 Try out this interactivity: Who is grabbing land? (int-3327)

5.7 SkillBuilder: Interpreting satellite images to show change over time

WHAT IS A SATELLITE IMAGE?

A satellite image is an image taken from a satellite orbiting the Earth. Satellite images allow us to see very large areas — much larger than those that can be visualised using vertical aerial photography.

Go online to access:

- a clear step-by-step explanation to help you master the skill
- a model of what you are aiming for
- a checklist of key aspects of the skill
- a series of questions to help you apply the skill and to check your understanding.

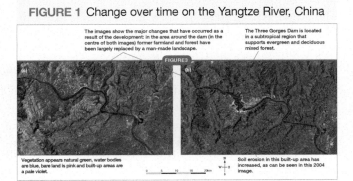

FIGURE 1 Change over time on the Yangtze River, China

The images show the major changes that have occurred as a result of the development: in the area around the dam (in the centre of both images) former farmland and forest have been largely replaced by a man-made landscape.

The Three Gorges Dam is located in a subtropical region that supports evergreen and deciduous mixed forest.

Vegetation appears natural green, water bodies are blue, bare land is pink and built-up areas are a pale violet.

Soil erosion in this built-up area has increased, as can be seen in this 2004 image.

5.8 Are we running dry?

5.8.1 Why are we running low on water?

There is no substitute for water. Without water there is no food, and agriculture already consumes 70 per cent of the world's fresh water. Every type of food production — cropping, grazing and processing — requires water. Thus, a lack of water is possibly the most limiting factor for increasing food production in future.

To feed an additional two billion people by 2050, the world will need to generate more food and use more water. The two main concerns that threaten future water security are water quantity and water quality (see figure 1).

In theory, the world has enough water; it is just not available where we want it or when we want it, and it is not easy to move from place to place. We already use the most accessible surface water, and now we are looking for it beneath our feet. Underground **aquifers** hold 100 times more water than surface rivers and lakes. However, groundwater is not always used at a sustainable rate, with extraction exceeding natural recharge, or filling. This occurs in many of the world's major food-producing places, in countries such as the United States, China and India.

FIGURE 1 Water scarcity is a serious threat to food security.

Water insecurity is connected with food insecurity. Figure 2 shows the predicted number of people who will face **water stress** and water scarcity in the future. A more complex view is seen in figure 3, which shows an interconnection between increased demand for water and predicted climate change, population increase and greater industrialisation in the 2050s.

FIGURE 2 People facing water stress and water scarcity

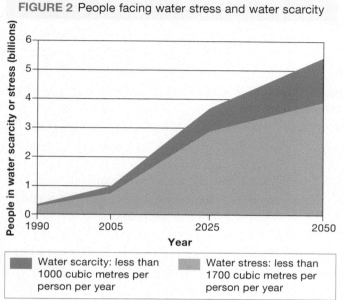

People in water scarcity or stress (billions)
6
5
4
3
2
1
0
1990
2005
2025
2050
Year
Water scarcity: less than 1000 cubic metres per person per year
Water stress: less than 1700 cubic metres per person per year

FIGURE 3 How water availability may change with temperature, population and industrialisation increase, 2050s

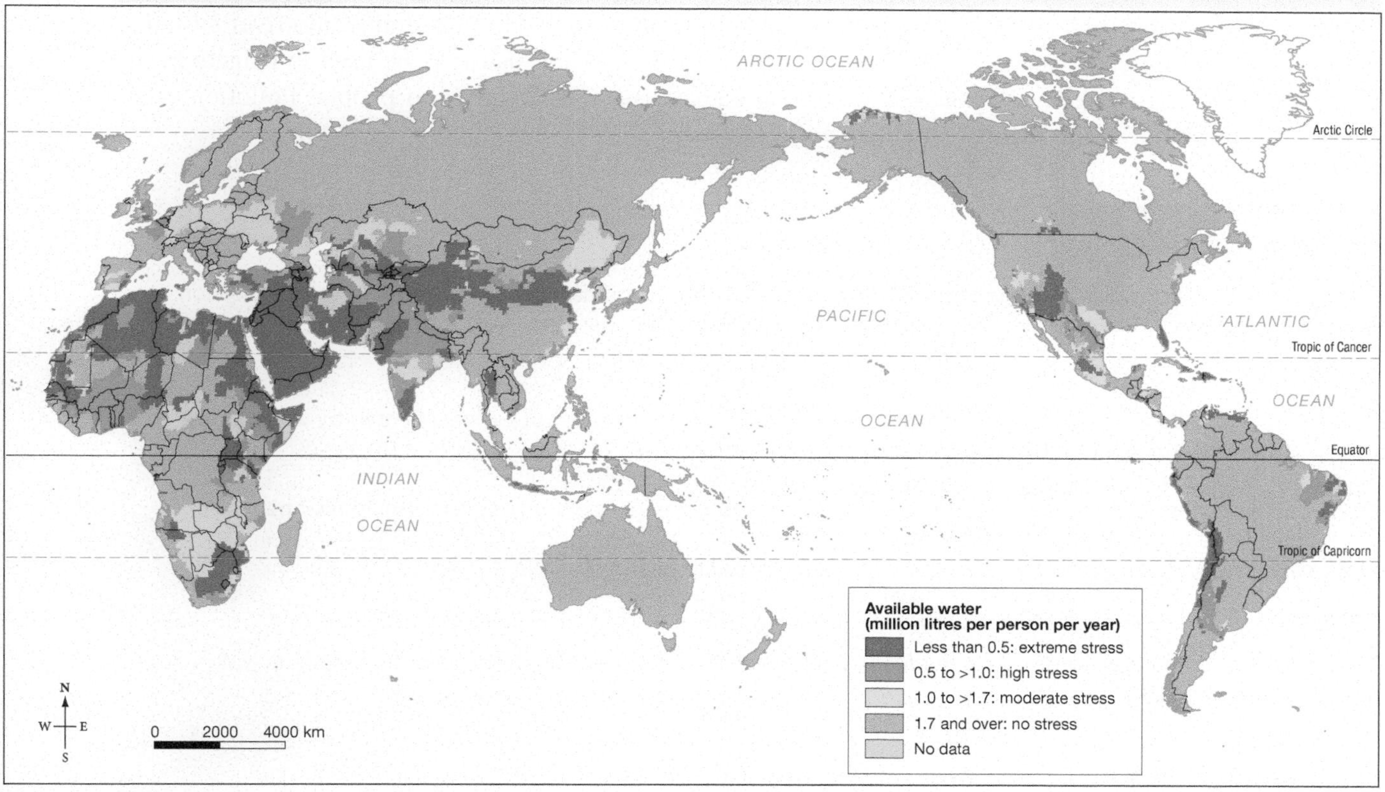

ARCTIC OCEAN
Arctic Circle
PACIFIC
OCEAN
ATLANTIC
OCEAN
Tropic of Cancer
Equator
INDIAN
OCEAN
Tropic of Capricorn
Available water
(million litres per person per year)
Less than 0.5: extreme stress
0.5 to >1.0: high stress
1.0 to >1.7: moderate stress
1.7 and over: no stress
No data
N
W
E
S
0
2000
4000 km

Source: Spatial Vision

When water availability drops below 1.5 million litres per person per year, a country needs to start importing food, although that makes the country susceptible to changes in global prices. Developing countries that experience water stress cannot afford to import food. They are also more vulnerable to environmental disasters. Seventy per cent of food emergencies in developing countries are brought on by drought.

The main causes of the growing water shortage are outlined below.

- *Food production*. It is estimated that an additional 6000 cubic kilometres of fresh water will be needed for irrigation to meet future food demand. Changes in diet, especially increased meat consumption, require more water to grow the crops and pasture that feed the animals. A typical meat eater's diet requires double the amount of water that a vegetarian diet requires.
- *Growth of urban and industrial demand*. Water for farming is diverted to urban populations, and productive land is converted to urban use.
- *Poor farming practices*. Water is wasted through inefficient irrigation methods and cultivating water-hungry crops such as rice. Poorly maintained irrigation infrastructure, such as pipes, canals and pumps, creates leakage.
- *Over-extraction*. Improved technology and cheaper, more available energy have enabled us to pump more groundwater from deeper aquifers. This is not always done at a sustainable rate, so as water is removed, less is available to refill lakes, rivers and wetlands.
- *Poor management*. Governments often price water cheaply, so irrigation schemes use water unsustainably. Some countries may have available water but lack the money to develop irrigation schemes.

5.8.2 Why is water quality deteriorating?

Agriculture is a major contributor to water pollution. Excess nutrients, pesticides, sediment and other pollutants can run off farmland or leach into soils and groundwater. Excessive irrigation can cause waterlogging or soil salinity. This salty water not only poisons the soil but also drains into river systems. Industrial waste, untreated sewage and urban run-off also pollute water that may be used to irrigate farmland. Food that is irrigated with polluted water can actually pass on diseases to people. Pollution is an important contributor to the scarcity of clean, **potable** water.

5.8 Activities

To answer questions online and to receive **immediate feedback** and **sample responses** for every question, go to your learnON title at www.jacplus.com.au. *Note*: Question numbers may vary slightly.

Remember

1. (a) Examine figure 2 and describe the projected *changes* in the number of people affected by water stress between 1990 and 2050. Use figures in your description.
   (b) How do these *changes* compare with figures for water scarcity?
2. If a country has an average of 0.5 to <1.0 million litres of water per person, per year, would they be considered to be water stressed? Why?
3. Why is agriculture both a contributor and a victim of water pollution?

Explain

4. (a) Refer to figure 3. Describe those *places* in the world that are predicted to be in high to extreme water stress in the 2050s.

(b) Compare your answer with a map of world average rainfall. Are areas that are predicted to be suffering high to extreme stress by 2050 also areas of low rainfall?

(c) How could you explain why *places* like eastern Europe could face water scarcity?

Discover

5. Use the **Water use** weblink in the Resources tab to select a country and find out more about its water usage. Using the data on this website, construct a table to compare water usage for four countries — one from each continent of Europe, Africa, Asia and South America. (Try to select different countries from those chosen by other students.) Write a paragraph to summarise your findings.

Predict

6. Use the **Water availability** weblink in the Resources tab and scroll to the 2020s map. Compare this with the map for 2050. What are the three most significant *changes* you can see?

Think

7. What do you think water managers could do to help prevent water scarcity affecting future food security?

learnon RESOURCES — ONLINE ONLY

Try out this interactivity: The last drop (int-3328)

Explore more with these weblinks: Water use, Water availability

5.9 Climate change: freeze or fry?

5.9.1 How will food security be affected by climate change?

The impacts of climate change on future world food security are a case of give and take. Some regions of the world will benefit from increases in temperature and rainfall, while others will face the threat of greater climatic uncertainty, lower rainfall and more frequent drought. In either case, food production will be affected.

Agriculture is important for food security, because it provides people with food to survive. It is also the main source of employment and income for 36 per cent of the world's workforce. In heavily populated countries in Asia, between 40 and 50 per cent of the workforce is engaged in food production, and this figure increases to over 63 per cent in sub-Saharan Africa.

FIGURE 1 Possible impacts of climate change on food production

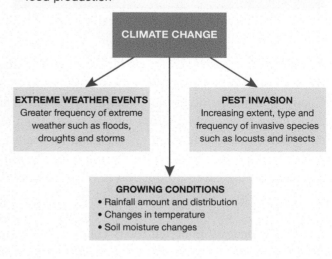

It is difficult to predict the likely impacts of climate change, because there are many environmental and human factors involved (see figure 1), as well as different predictions from scientists (see figure 2). Use the **How to feed the world in 2050** weblink in the Resources tab to find out more about this topic.

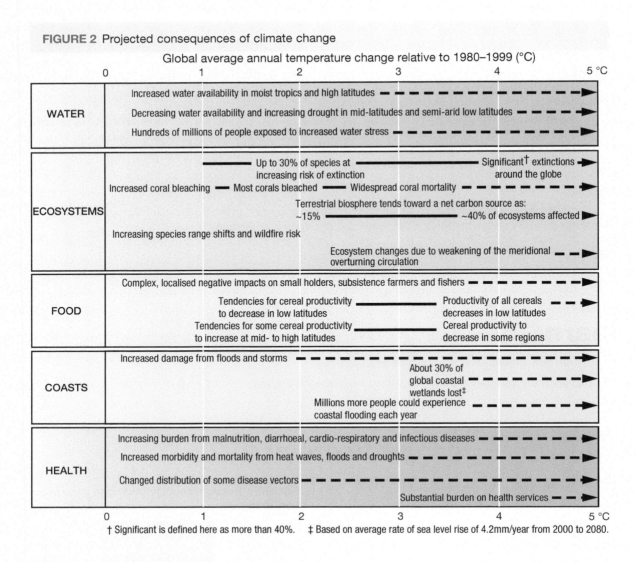

FIGURE 2 Projected consequences of climate change

Global average annual temperature change relative to 1980–1999 (°C)

WATER
- Increased water availability in moist tropics and high latitudes
- Decreasing water availability and increasing drought in mid-latitudes and semi-arid low latitudes
- Hundreds of millions of people exposed to increased water stress

ECOSYSTEMS
- Up to 30% of species at increasing risk of extinction ——— Significant† extinctions around the globe
- Increased coral bleaching — Most corals bleached — Widespread coral mortality
- Terrestrial biosphere tends toward a net carbon source as: ~15% ——— ~40% of ecosystems affected
- Increasing species range shifts and wildfire risk
- Ecosystem changes due to weakening of the meridional overturning circulation

FOOD
- Complex, localised negative impacts on small holders, subsistence farmers and fishers
- Tendencies for cereal productivity to decrease in low latitudes ——— Productivity of all cereals decreases in low latitudes
- Tendencies for some cereal productivity to increase at mid- to high latitudes ——— Cereal productivity to decrease in some regions

COASTS
- Increased damage from floods and storms
- About 30% of global coastal wetlands lost‡
- Millions more people could experience coastal flooding each year

HEALTH
- Increasing burden from malnutrition, diarrhoeal, cardio-respiratory and infectious diseases
- Increased morbidity and mortality from heat waves, floods and droughts
- Changed distribution of some disease vectors
- Substantial burden on health services

† Significant is defined here as more than 40%.   ‡ Based on average rate of sea level rise of 4.2mm/year from 2000 to 2080.

There is a wide range of possible impacts of climate change. Sea-level rises may cause flooding and the loss of productive land in low-lying coastal areas, such as the Bangladesh and Nile River deltas. Changes in temperatures and rainfall may cause an increase in pests and plant diseases. However, agriculture is adaptable. Crops can be planted and harvested at different times, and new types of seeds and plants, or more tolerant species, can be used. Low-lying land may be lost, but higher elevations, such as mountain slopes, may become more suitable. The loss in productivity in some places may be balanced by increased production in other places. Figure 3 demonstrates the effects of climate change on cereal crops, while figure 4 shows the range of potential impacts across the European Union.

Essentially, hundreds of millions of people are at risk of increased food insecurity if they have to become more dependent on imported food. This will be evident in the poorer countries of Asia and sub-Saharan Africa, where agriculture dominates their economy. There is also a risk of greater numbers of **environmental refugees** or people fleeing places of food insecurity.

FIGURE 3 Predictions of the effects of climate change on cereal crops

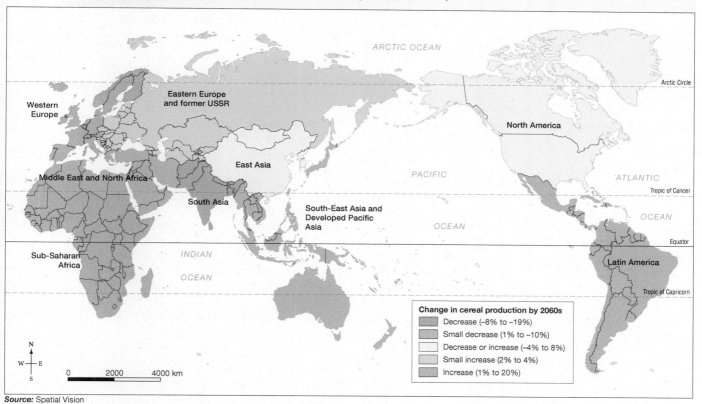

Change in cereal production by 2060s
- Decrease (−8% to −19%)
- Small decrease (1% to −10%)
- Decrease or increase (−4% to 8%)
- Small increase (2% to 4%)
- Increase (1% to 20%)

Source: Spatial Vision

FIGURE 4 Examples of potential consequences of climate change in the European Union

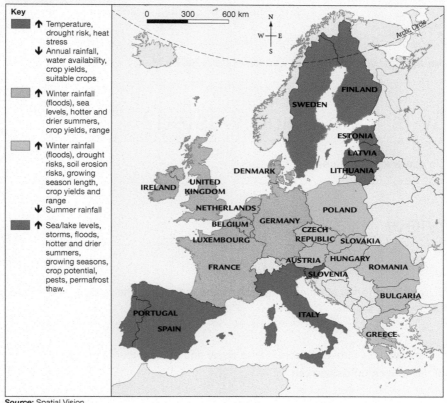

Key
- ↑ Temperature, drought risk, heat stress
  ↓ Annual rainfall, water availability, crop yields, suitable crops
- ↑ Winter rainfall (floods), sea levels, hotter and drier summers, crop yields, range
- ↑ Winter rainfall (floods), drought risks, soil erosion risks, growing season length, crop yields and range
  ↓ Summer rainfall
- ↑ Sea/lake levels, storms, floods, hotter and drier summers, growing seasons, crop potential, pests, permafrost thaw.

Source: Spatial Vision

5.9 Activities

To answer questions online and to receive **immediate feedback** and **sample responses** for every question, go to your learnON title at www.jacplus.com.au. *Note*: Question numbers may vary slightly.

Remember

1. Refer to figure 2 and decide whether the following statements are true or false.
   (a) If temperatures increase by 3 °C, you would expect to see crop yields rising around the equator.
   (b) *Changes* in extreme weather events are unlikely unless temperatures increase by at least 1 °C.
   (c) Food insecurity will be felt greatly in developing regions if temperatures rise more than 4 °C.
   (d) *Places* that are likely to experience decreasing crop yields will be found in the higher latitudes.

Explain

2. (a) Refer to figure 3. Which *places* have the potential to be grain exporters and which *places* are likely to become dependent on grain imports? Use data in your answer.
   (b) What are the economic and social implications of this for countries in these regions?
3. (a) Refer to figure 4. Which countries of Europe will benefit from climate *change* in terms of food production and which countries are likely to suffer negative outcomes?
   (b) Would increased irrigation be a *sustainable* solution to growing food in Spain? Explain your answer.
4. Describe the *interconnection* between *environmental* refugees and climate *change*.

Discover

5. Research potential impacts of climate *change* on Australia. Create an annotated map to illustrate your findings.

Predict

6. How might a country such as Australia best prepare its food production systems to cope with potential *changes* in climate?

Think

7. How might food be shared more equitably around the world? Discuss with a group and report your suggestions back to the class.

 **learn** **on** RESOURCES — ONLINE ONLY

 Explore more with this weblink: How to feed the world in 2050

5.10 Why is food being wasted?

Access this subtopic at **www.jacplus.com.au**

5.11 Review

5.11.1 Review

The Review section contains a range of different questions and activities to help you revise and recall what you have learned, especially prior to a topic test.

5.11.2 Reflect

The Reflect section provides you with an opportunity to apply and extend your learning.

Access this subtopic at **www.jacplus.com.au**

TOPIC 6
2050 — food shortage or surplus?

6.1 Overview

Numerous **videos** and **interactivities** are embedded just where you need them, at the point of learning, in your learnON title at www.jacplus.com.au. They will help you to learn the content and concepts covered in this topic.

6.1.1 Introduction

Currently we produce enough food to adequately feed everyone in the world. However, it is estimated that in 2015 about 795 million people or almost one in seven went hungry. The world's population is expected to grow by another two billion people in the next 30 years. If we want to stop the number of hungry people from increasing, we will need improvements in food production, new sources of food, better aid programs, and different attitudes to food consumption and waste.

- Twenty-five per cent of people in sub-Saharan Africa are undernourished.
- Use of fertilisers will expand in Latin America, East Asia, North America and South Asia.
- We can act to decrease hunger in developed and less developed regions.
- Could Australia become the food bowl of Asia?
- Should farmers be stewards of the environment?
- Chinese people now eat a quarter of the world's supply of meat.
- Poor nutrition causes 45 per cent of all deaths in children under five in developing nations.
- Home-grown vegetables do not necessarily have a smaller carbon footprint than those from a supermarket.
- Biofuel opportunities will mostly occur in developing regions, providing many jobs.
- Can we feed the 9.5 billion people expected to live on the planet in 2050?

INQUIRY SEQUENCE

6.1	Overview	107
6.2	Can we feed the future world population?	108
6.3	Can we improve food production?	111
6.4	What food aid occurs at a global scale?	115
6.5	**SkillBuilder:** Constructing a box scattergram	118
6.6	Do Australians need food aid?	118
6.7	Is trade fair?	121
6.8	**SkillBuilder:** Constructing and describing proportional circles on maps	123
6.9	How do dietary changes affect food supply?	124
6.10	Can urban farms feed people?	126
6.11	**Review**	129

6.2 Can we feed the future world population?

6.2.1 What is the problem of hunger?

One in seven people in the world does not have enough food to lead an active and healthy life, yet over one billion people are overweight. How can we best manage the challenge of ensuring that everyone in the world has access to a healthy and adequate diet?

6.2.2 What are the impacts of hunger?

The distribution of the world's population and the availability of arable land per person is uneven. Regions with the fastest growing future populations are also those where there is limited arable land per person (see figure 1).

The impact of hunger on people cannot be overstated. Hunger kills more people each year than disease (see figure 2). It is estimated that we will need to produce between 70 and 100 per cent more food in order to feed future populations. New ideas, knowledge and techniques will be needed if we do not want millions more people to suffer malnourishment, starvation and vulnerability to disease. The challenge, though, is to do this in a way that is also sustainable. However, population growth and limited supplies of arable land will affect how much food can be produced.

One solution to feeding people who live in crowded spaces, such as Asia, or in environmentally challenging spaces, such as sub-Saharan Africa, is to increase the amount of trade in food products. This will involve moving food from places with crop surpluses (North America, Australia and Europe) to regions that are crowded or less productive. This means there will be an increase in the interconnection between some countries.

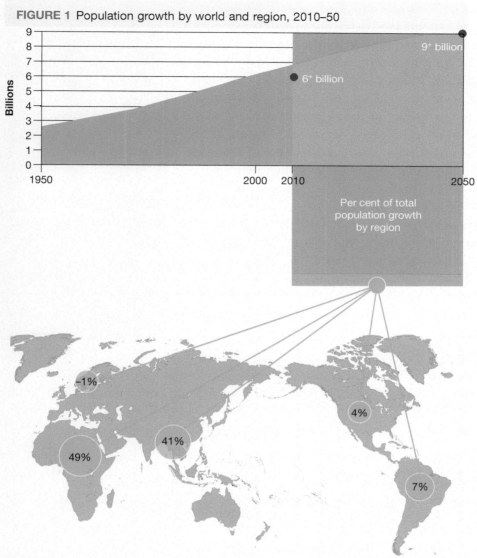

FIGURE 1 Population growth by world and region, 2010–50

9⁺ billion

6⁺ billion

Per cent of total population growth by region

−1%

4%

41%

49%

7%

Source: Re-drawn from an image by Global Harvest Initiative (*2011 GAP Report®: Measuring Global Agricultural Productivity*), data from the United Nations.

FIGURE 2 Hunger is the world's number one health risk, killing more people each year than AIDS, malaria and tuberculosis combined.

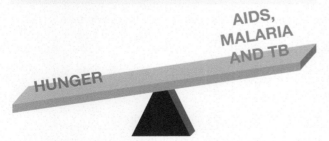

AIDS, MALARIA AND TB

HUNGER

6.2.3 Plumpy'nut — a short-term solution to malnutrition

In 2005 a revolutionary approach to treating malnutrition was released. This is a ready-to-use therapeutic food (RUTF) called Plumpy'nut. It is a sweet, edible paste made of peanut butter, vegetable oils, powdered milk, sugar, vitamins and minerals. Its advantages are that it:

• is easy to prepare
• is cheap (a sachet costs about $2.50)
• needs no cooking, refrigeration or added water
• has a shelf life of two years.

Children suffering from malnutrition can be fed at home without having to go to hospital. It is specially formulated to help malnourished children regain body weight quickly, because malnutrition leads to stunting of growth, brain impairment, frailty and attention deficit disorder in children under two years of age.

Plumpy'nut is not a miracle cure for hunger or for malnutrition; it only treats extreme food deprivation, mainly associated with famines and conflicts. It is not designed to reduce chronic hunger resulting from long-term poor diets or malnutrition. Since its introduction, Plumpy'nut has lowered mortality rates during famines in Malawi, Niger and Somalia. Most of the world's peanuts are grown in developing countries, where allergies to them are relatively uncommon. Manufacturing plants have been established in a dozen developing countries, including Mali, Niger and Ethiopia. These factories provide employment and ensure ease of access when needed. The patent for Plumpy'nut is owned by Nutriset, a French company. There are some concerns that this gives Nutriset the power to entirely control the production and distribution of Plumpy'nut. Nutriset has been asked by Médecins sans Frontières (Doctors without Borders) to remove its patent, which would allow open production and distribution of this life-saving product.

FIGURE 3 Plumpy'nut benefits children.

6.2 Activities

To answer questions online and to receive **immediate feedback** and **sample responses** for every question, go to your learnON title at www.jacplus.com.au. *Note:* Question numbers may vary slightly.

Remember

1. Examine figure 1.
   (a) Which region is predicted to decrease in population by 2050?
   (b) Which two continents are expected to have the greatest increase in population?
   (c) What is the predicted world population in 2050?
   (d) How does a growing world population put pressure on food shortages?

Explain

2. Explain why you think hunger may threaten people's health and be responsible for so many deaths.
3. What other geographic factors can cause shortages of food?
4. Outline the advantages and disadvantages of using Plumpy'nut or other RUTFs to treat childhood malnutrition in developing countries.
5. What may need to happen to ensure there is enough food in the future for people who live in *places* with growing populations and limited arable land?

Discover

6. As well as affecting people's health, a shortage of food can have social and political effects. Undertake research into the series of food riots that occurred in a number of countries around the world in 2015.
   (a) Where did these riots occur?
   (b) What were the causes of these riots?
   (c) Why might governments need to prevent this situation from occurring again?
7. What foods are commonly used as food aid worldwide?

Predict

8. Lack of food has been a factor in pushing people to leave their homes and go to cities in search of employment and food. Predict the *places* of the world where this is most likely to happen.

Think

9. Draw a poster or advertisement that informs Australians about Plumpy'nut and seeks donations for its use.

6.3 Can we improve food production?

6.3.1 How can we improve food production?

Over 70 per cent of the world's poor live in rural areas, and improving their lives would create greater food security. If poor farmers can produce more food, they can feed themselves and provide for local markets. Improved infrastructure, such as roads in rural regions, would enable them to transport their produce to market and increase their incomes. Preventing hunger on a global scale is important, but action is also needed on a local scale.

Future changes to food production

Farming is a complex activity, and farmers around the world face many challenges in producing enough food to feed themselves and to create surpluses they can sell to increase their incomes. As urban areas grow, the amount of arable land available decreases. According to the United Nations Food and Agriculture Organization (FAO), the world has an extra 2.8 billion hectares of unused potential farmland. This is almost twice as much as is currently farmed. However, only a fraction of this extra land is realistically available for agricultural expansion, owing to inaccessibility and the need to preserve forest cover and land for infrastructure.

FIGURE 1 Factors affecting farming yields

Availability of surface water and groundwater

Impact of natural disasters: floods, storms, drought

Length of growing season

Government regulations and policies

Climate: rainfall amounts, seasonality and temperature

Amount of money spent on agricultural research

Availability of insects for pollination

Impact of insects and diseases

Soil fertility and type

Access to finance: micro-loans

Ease of access to markets

In Australia there are hopes to expand the agricultural output from the Ord River irrigation scheme in the Kimberley region of Western Australia.

The growing populations of the future will be found in places where expansion of agricultural land is already limited. Consequently, increased food production will need to come from better use of current agricultural areas, better use of technology, and new ways of thinking about food production and approaches to farming. Figure 2 outlines some strategies that may be used to improve food production.

FIGURE 2 Strategies for improving sustainable food production

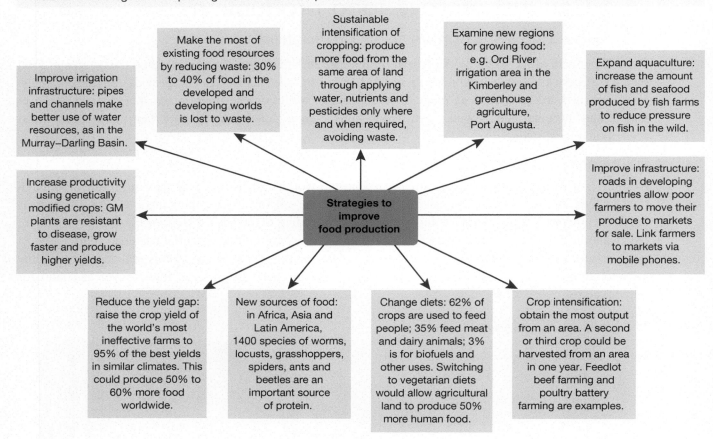

Improve irrigation infrastructure: pipes and channels make better use of water resources, as in the Murray–Darling Basin.

Make the most of existing food resources by reducing waste: 30% to 40% of food in the developed and developing worlds is lost to waste.

Sustainable intensification of cropping: produce more food from the same area of land through applying water, nutrients and pesticides only where and when required, avoiding waste.

Examine new regions for growing food: e.g. Ord River irrigation area in the Kimberley and greenhouse agriculture, Port Augusta.

Expand aquaculture: increase the amount of fish and seafood produced by fish farms to reduce pressure on fish in the wild.

Increase productivity using genetically modified crops: GM plants are resistant to disease, grow faster and produce higher yields.

Strategies to improve food production

Improve infrastructure: roads in developing countries allow poor farmers to move their produce to markets for sale. Link farmers to markets via mobile phones.

Reduce the yield gap: raise the crop yield of the world's most ineffective farms to 95% of the best yields in similar climates. This could produce 50% to 60% more food worldwide.

New sources of food: in Africa, Asia and Latin America, 1400 species of worms, locusts, grasshoppers, spiders, ants and beetles are an important source of protein.

Change diets: 62% of crops are used to feed people; 35% feed meat and dairy animals; 3% is for biofuels and other uses. Switching to vegetarian diets would allow agricultural land to produce 50% more human food.

Crop intensification: obtain the most output from an area. A second or third crop could be harvested from an area in one year. Feedlot beef farming and poultry battery farming are examples.

Strategies for improving food production

The strategy that is likely to be the most important in increasing future crop production is called closing the **yield gap**. This means that farmers who are currently less productive will need to increase their yields so that their outputs are closer to those of the more productive farmers. There is a serious yield gap in more than 157 countries. If this were achieved, larger amounts of food would be available without needing more land. There are wide geographic variations in crop and livestock productivity. Brazil, Indonesia, China and India have all made great progress in increasing their agricultural output. Much of the increase has been achieved through more efficient use of water and fertilisers.

The use of **genetically modified** (GM) foods has increased, and this has increased crop yields. However, there is some opposition to GM crops because of concerns about:
- their safety
- loss of seed varieties
- potential risks to the environment and people's health
- the fact that large companies hold the copyright to the seeds of GM plants that are food sources.

6.3.2 Solving problems and coming up with solutions

Because agriculture uses 60 to 80 per cent of the planet's increasingly scarce fresh water resources, any method that can produce food without needing fresh water at all is a great advance.

Port Augusta is located in a hot, arid region of South Australia, and is not normally associated with agriculture. However, one company, Sundrop Farms, is using this region's abundant renewable resources of sunlight and sea water to produce high-quality, pesticide-free vegetables, including tomatoes, capsicums and cucumbers, and it does so all year round.

Large mirrors concentrate the sun's energy and the collected heat creates steam to drive electricity production, heat the greenhouse, and desalinate sea water, producing up to one million litres of fresh water a day for crop irrigation.

Experimental greenhouse in Port Augusta

The current experimental greenhouse, covering 2000 square metres, is operating on a small scale. A new $150 million greenhouse will be completed approximately 10 kilometres from the site in 2017. This new project aims to produce approximately 15 000 tonnes of truss tomatoes per year, which equates to about 10 per cent of Australia's demand. The new greenhouse uses renewable energy to be not just carbon and water neutral, but positive. It is hoped that this type of technology can be used in many more places in Australia and around the world that have hot, arid climates previously unsuitable for horticulture. The technology has the potential to supply millions of people with healthy food in a sustainable manner while also using limited fossil fuel resources.

Australian farmers see technology as a means of decreasing production costs and increasing crop production. Additional technologies in Australian agriculture include the following.

- Robots are being tested to determine whether they can be used in complex jobs such as watering or harvesting. This would be of advantage in the horticultural sector, which is the third largest sector in agriculture, with an export trade worth $2.1 billion in 2014–15.
- Technology such as satellite positioning is being used to determine the optimal amounts of fertiliser to use on crop farms, which could increase profitability by as much as 14 per cent.
- Robots and an unmanned air vehicle have passed field tests at an almond farm in Mildura, Victoria. They are fitted with sensors, vision, laser, radar and conductivity sensors — including GPS and thermal sensors.

FIGURE 3 Yield gap for a combination of major crops, 2015

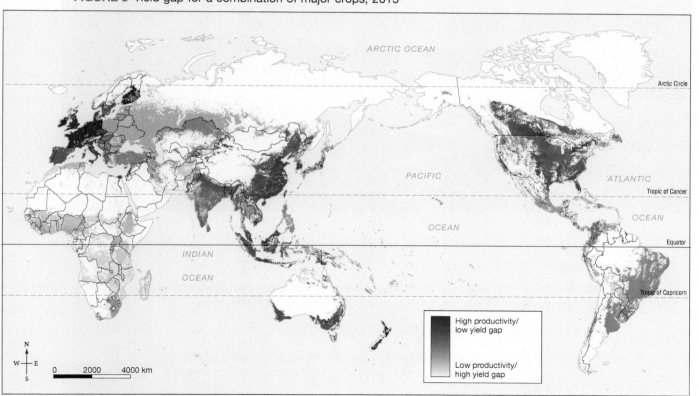

Source: Food and Agriculture Organization of the United Nations

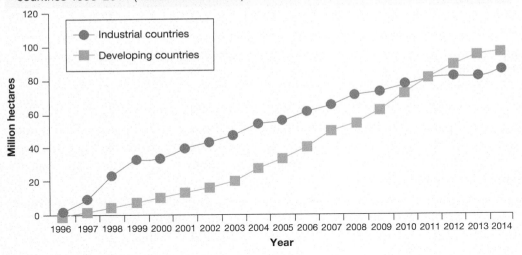

FIGURE 4 Global area of genetically modified crops in industrial and developing countries 1996–2014 (millions of hectares)

FIGURE 5 The world's first Sundrop Farm is situated in Port Augusta, South Australia

6.3 Activities

To answer questions online and to receive **immediate feedback** and **sample responses** for every question, go to your learnON title at www.jacplus.com.au. *Note:* Question numbers may vary slightly.

Remember

1. List three different strategies, other than closing the yield gap, for improving food production.
2. What is meant by the term *yield gap* and why is it important that this gap be narrowed to increase future crop yields? See figure 3.

3. What is meant by the term *genetically modified* (GM)?
4. Examine figure 4. What *changes* have there been in the production of genetically modified foods in (a) developed countries and (b) developing countries? Use the data from the graph to support your answer.
5. (a) Referring to figure 3, which *places* have the highest gap yields?
   (b) Which places in figure 3 have the lowest gap yields?

Explain

6. Explain the advantages and disadvantages of locating the large greenhouse near Port Augusta. You may need to consult your atlas.
7. Select one of the strategies outlined in figure 2 that can be used to improve food production. Explain this strategy in your own words and outline some of the strengths and weaknesses of this strategy.
8. Use the **Vertical farming** weblink in the Resources tab to watch a video clip on this topic. What is being suggested about *environmentally sustainable* farming in the future? Draw a diagram to show what a future vertical farm might look like. How might a vertical farm help in feeding future populations?

Predict

9. Many Australian cities have large housing estates on their outskirts. This land was often used for market gardens or farmland. What impact might the loss of this productive land have on the price of food?
10. Predict what the impact might be on people and *places* if the greenhouse method of farming shown in figure 5 were to become more readily available. What might be the effects on *places* where the yield gap is large compared to *places* that are currently more productive?

Think

11. Annotate key aspects of the Port Augusta landscape (greenhouses, solar collector, sandy soil, flat and barren landscape) shown in the photograph in figure 5.
12. Some areas of Australia that are currently national parks or marine parks may be sought after as agricultural land in future. Outline your views on this.

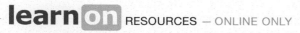 **learn** on RESOURCES — ONLINE ONLY

 Try out this interactivity: More, or less, food (int-3329)

 Explore more with this weblink: Vertical farming

6.4 What food aid occurs at a global scale?

6.4.1 Who needs food aid and how is it delivered?

The World Food Programme (WFP) is a voluntary arm of the United Nations. It reaches more than 80 million people, in more than 92 countries, with food assistance after disasters and conflicts. Food aid is food, money, goods and services given by wealthier, more developed nations to less developed nations for the specific purpose of helping the poor.

FIGURE 1 How the WFP works

People who need food aid include:

- poor people who cannot buy food even if it is available, as they are often trapped in a cycle of hunger and poverty
- people who have fled violence or civil conflict
- people devastated by natural disasters.

The WFP provides different types of food aid to people after natural disasters such as cyclones, floods and earthquakes. Some relief aid is provided in the short term as emergency food. Project food relief is often required over lengthy periods, typically after civil war or prolonged drought.

Project food aid is geared towards future disaster prevention. Program food aid is organised between national governments.

6.4.2 Who gives food aid?

The major donor countries to the WFP in 2015 are shown in table 1.

TABLE 1 Major funding contributors to WFP in 2015 (US$)

All donors & funding sources		
1	USA	2 008 800 185
2	United Kingdom	456 855 096
3	Germany	329 258 331
4	Canada	261 645 796
5	European Commission	250 347 378
6	Japan	196 773 084
7	UN CERF	159 928 948
8	Saudi Arabia	151 249 675
9	Netherlands	101 464 033
10	Private Donors	98 332 463

Figures current as at 24 July 2016

6.4.3 Who receives food aid?

In 2014–15, the WFP worked on six simultaneous Level 3 emergencies including providing food assistance to 2.2 million people affected by conflict in Iraq; more than one million people affected by Typhoon Haiyan in the Philippines; over two million people in South Sudan; more than million people disrupted by the Ebola outbreak in Guinea, Sierra Leone and Liberia; and over 4.9 million people displaced by war in Syria. The WFP directly helped over 66.8 million women and children, 14.8 million internally displaced people and over 68 million refugees. In total, the WFP provided more than 3.3 million tonnes of food aid.

CASE STUDY

Cash vouchers and school feeding programs

Where food is available but people simply cannot afford to buy it, aid in the form of cash vouchers from the WFP can be exchanged for food and other essential commodities. They allow recipients greater choice in the types of food and other commodities they obtain. Cash has benefits for local economies because the money is spent within the community. Recently, cash voucher programs have been enhanced through the use of mobile phones, which have been used to provide instant payments to both beneficiaries and the shopkeepers who redeem vouchers. In 2014–15, the WFP provided school meals to over 17 million children.

Another program provides school children with either full meals (breakfast and/or lunch) or nutritional snacks, such as high-energy biscuits. In some cases, school meals are provided alongside take-home rations that benefit the whole family and provide an additional incentive for sending children to school.

In 2015, Australia funded school feeding programs in Bangladesh, Myanmar, Laos and Cambodia. These have had strong positive impacts on both the children and the wider community. School rates of enrolment increased and regular attendance improved. Households also benefited through a reduced need to purchase food. Australia will contribute more than $167.5 million of funding through the World Food Programme from 2015 to 2020. This includes funding provided for the food crises after Tropical Cyclone Pam in Vanuatu and the Nepal earthquake.

6.4 Activities

To answer questions online and to receive **immediate feedback** and **sample responses** for every question, go to your learnON title at www.jacplus.com.au. *Note:* Question numbers may vary slightly.

Remember

1. Refer to figure 1. How many countries received food aid in 2014?
2. Refer to table 1. How much food aid did Japan give to the WFP in 2015?
3. Refer to figure 1. Which regions of the world receive the greatest food aid? Suggest reasons for this.

Explain

4. Explain why the WFP is so active in school feeding and emergency aid programs.
5. Refer to the case study 'Cash vouchers and school feeding programs'.
   (a) Locate the recipient countries on a world map.
   (b) List the advantages and disadvantages of cash vouchers.
   (c) Use the **World Food Programme** weblink in the Resources tab to help you explain the benefits of food aid shown here.

Discover

6. Select a major donor of food aid from table 1. What are the main population characteristics of this country, such as life expectancy, literacy levels and death rates? Discuss your findings in class.
7. Use the **World Food Programme** weblink in the Resources tab to help you explain WFP's involvement in Syria and surrounding *places* since 2012. What action is the WFP taking there and why?

Predict

8. How might food aid *change* when a donor country experiences a major economic downturn?
9. Predict the likely consequences for children who suffer from malnutrition. Present your information in an appropriate diagram.

Think

10. Should Australia's food aid commitment be increased? Write a letter to your federal member of parliament, outlining your views on increasing Australia's food aid contribution.

 RESOURCES — ONLINE ONLY

🔗 **Explore more with this weblink:** World Food Programme

6.5 SkillBuilder: Constructing a box scattergram

WHAT IS A BOX SCATTERGRAM?

A box scattergram is a table with columns and rows that displays the relationship between two sets of data that have been mapped. Box scatter-grams are a useful way of summarising data from maps.

Go online to access:

- a clear step-by-step explanation to help you master the skill
- a model of what you are aiming for
- a checklist of key aspects of the skill
- a series of questions to help you apply the skill and to check your understanding.

FIGURE 1 A box scattergram showing the relationship between undernourishment and aid received per person

Hunger level (% undernourished)	Aid received per person (US$)			
	No data	Less than 20	20–99	Over 100
35 +				• Congo • Mozambique
25–34			• Chad • Angola	
15–24				
5–14		• Nigeria	• Niger	
Less than 5		• South Africa • Algeria • Libya		
No data				• Mauritania

learn on RESOURCES — ONLINE ONLY

🎞 **Watch this eLesson:** Constructing a box scattergram (eles-1734)

💠 **Try out this interactivity:** Constructing a box scattergram (int-3352)

6.6 Do Australians need food aid?

6.6.1 Who needs food aid in Australia?

In 2015, it was estimated that about 2.5 million, or 13.9 per cent, of Australians are living below the inter-nationally accepted poverty line. This includes almost 603 000 children or 17.7 per cent of the total Australian child population. The prices of essentials — food, health, education, housing, utilities and transport — have climbed so much in recent years that people who are already struggling are unable to cope. They may need food aid. The economic climate saw people turning to charity who would never have dreamed of seeking such support in the past. It is not just traditionally vulnerable groups, such as the home-less, who are seeking food relief; it is also the aged, single parents and the 'working poor'.

In 2014 it was reported that:

- one in 10 Australians on unemployment benefits was unable to obtain a substantial meal each day
- two million people across the country could not secure enough food for their families
- nine out of 10 welfare agencies, from 600 interviewed, claimed they could not meet the demand for food, and there had been an 11 per cent increase in welfare agencies seeking food
- 105 000 Australians were homeless
- between 2010 and 2012, poverty increased by over 1 per cent
- 40.1 per cent of people on social security payments were living below the poverty line
- one in four pensioners lived close to the poverty line
- food banks provided enough food for 32 million meals per year.

Redistributing food through collection and food agencies is a good way to make better use of food resources.

The services of groups such as Meals on Wheels may be in greater demand as Australia's population ages (see figures 1 and 2). In 1997, the **median age** was 34 years, but this is projected to be 44–46 years in 2051. In 1997, people aged 65 years and over comprised 12 per cent of the population, and this is projected to rise to 24–26 per cent in 2051.

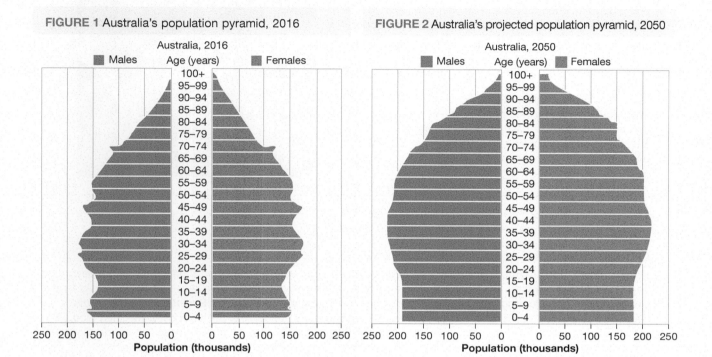

FIGURE 1 Australia's population pyramid, 2016

FIGURE 2 Australia's projected population pyramid, 2050

CASE STUDY 1

SecondBite

SecondBite rescues and redistributes food to agencies that service people in need. Food is donated from the farm gate, wholesalers, markets, supermarkets and caterers.

SecondBite has rescued more than 20 million kilograms of food nationally and redistributed this food free of charge to more than 1200 community food program partners.

FIGURE 3 A SecondBite delivery truck, which is used to deliver donated food to people in need

CASE STUDY 2

Meals on Wheels

Meals on Wheels began in the United Kingdom during World War II. In Australia, it began in Melbourne in 1952, and it plays an important role in helping aged and disabled people to remain in their homes. Some people may not be able to get out and about, so over 78 000 volunteers help to deliver meals to them. By providing nutritious, relatively cheap meals, costing between $4.50 and $9.00, Meals on Wheels helps to make it possible for people to maintain their independence. Some 14.8 million meals are served annually to 53 000 people across the nation. The social interaction and regular visits are an important part of this service.

FIGURE 4 Volunteers for Meals on Wheels

CASE STUDY 3

Outback Australia

In September 2011, Indigenous Australian peoples in South Australia were suffering food shortages because they could not afford to buy food. The Red Cross and the South Australian Government were forced to send pallets of food to impoverished people living in the Anangu Pitjantjatjara Yankunytjatjara (APY) lands in South Australia's far north. Shops in the APY lands were full of food, but people were going hungry because of the high cost of freighted food. In June 2010, essential foods in remote community stores were more than double the price of those in Adelaide. One suggested strategy for improving food security in such places is the establishment of community gardens.

APY lands program

Since 2014, the APY lands program has been funded solely by non-government organisations (NGOs). The program now focuses on improving freight efficiency, stores management, cold storage upgrades and the supply of more generators for a reliable source of power.

6.6 Activities

To answer questions online and to receive **immediate feedback** and **sample responses** for every question, go to your learnON title at www.jacplus.com.au. *Note:* Question numbers may vary slightly.

Remember

1. Examine figures 1 and 2.
   (a) How many Australians were over 65 years of age in 2015?
   (b) How many are expected to be over 65 years in 2050?
2. How many people were living in poverty in Australia in 2014?
3. How many meals are served each year by SecondBite?

Explain

4. Explain why there might be difficulties with access to food in 2050 if 25 per cent of the population is over 65.
5. Explain the importance of volunteers in food redistribution.
6. Explain how Australia's size could lead to food shortage in some *places*.

Discover

7. Conduct your own research into a local organisation that provides food aid. Present your findings in an appropriate format.
8. Use the **Australian poverty** weblink in the Resources tab to discover other aspects of poverty in Australia.

Predict

9. Predict whether Meals on Wheels will experience an increase or decrease in its future clientele. Apart from the ageing population, what other factors might *change* the demand for SecondBite or Meals on Wheels in future?
10. What would be your family's reaction if the cost of food doubled because of freight costs? What steps could improve the situation in outback areas?

Think

11. Have your attitudes to redistributing food *changed* as a result of your reading and class discussion?
12. 'When bills have to be paid, food becomes a **discretionary item**' (Food Bank Australia 2011). If household bills have to be paid before buying food, what are the likely consequences for families and organisations supplying food aid?

 **learn on** RESOURCES — ONLINE ONLY

 Explore more with this weblink: Australian poverty

6.7 Is trade fair?

6.7.1 Why is trade not fair for everyone?

As people become more concerned about the level of poverty and hunger in the world, they sometimes seek ways to improve the situation. Trade is the way countries sell what they have produced and buy what they need. On a global scale, this does not mean that trade is mutually beneficial. Trade usually favours countries that are the strongest economically, and it disadvantages those countries that are poor.

Trade is not a level playing field. It favours the strongest countries, often to the disadvantage of the poor and weak. Strong or developed countries are able to:

- stockpile or dump crops so that they sell for a maximum price
- negotiate some political advantage
- refuse to sell crops if they have a shortage, so some nations go hungry
- dump crops of low quality onto poorer nations.

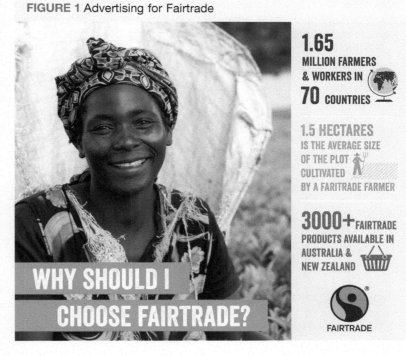

FIGURE 1 Advertising for Fairtrade

1.65 MILLION FARMERS & WORKERS IN **70** COUNTRIES

1.5 HECTARES IS THE AVERAGE SIZE OF THE PLOT CULTIVATED BY A FAIRTRADE FARMER

3000+ FAIRTRADE PRODUCTS AVAILABLE IN AUSTRALIA & NEW ZEALAND

FAIRTRADE

WHY SHOULD I CHOOSE FAIRTRADE?

In contrast, fair trade is a consumer-driven movement to promote fair prices and reasonable conditions for producers in developing regions. It tries to ensure, for example, that a group of farmers is able to sell its crop for a guaranteed price. Fair trade is seen by some as **active consumerism**. In addition to the general movement known as fair trade, there is an organisation called Fairtrade International. This is a group of 25 organisations trying to ensure a better deal for producers.

6.7.2 Why do small farmers turn to Fairtrade certification?

In Ghana, cocoa farmers with 2- to 3-hectare plots of land face a number of problems.

- They may grow one main cash crop for low prices that do not even cover the costs of production; they may also grow some vegetables for family use and extra income.
- They face expensive production inputs (tools, fertilisers and pesticides) and family costs (education, medicine, food and clothes).
- Poor education makes it difficult for them to try different crops that could earn them more money.
- The use of child and slave labour is common in cocoa-producing West African countries.
- Technical, financial and scientific advice is limited.
- Credit is a key issue for farmers with **seasonal crops** like cocoa. Farmers need money for food and immediate needs, as well as pre-financing for planting and cultivation of their crop. Most borrow money from family, and a small percentage take a loan.
- There are taxes or tariffs on products.

As a consequence, many farmers are turning to Fairtrade as a means of improving their livelihoods.

Where are Fairtrade farmers found?

Countries with Fairtrade organisations are in the poorer nations. They are frequently found in Asia, Africa and South America (see figure 2). These farmers produce crops such as grapes, cocoa, coffee, tea, dried fruits, bananas, sugar, rice, nuts and handicrafts. The consumers who are likely to buy crops produced by Fairtrade farmers are located in the wealthy nations, because they have the ability to pay a little more to buy their goods.

FIGURE 2 Map showing Fairtrade producer countries. The annotations describe how selected producers use their Fairtrade premium — a sum of money paid in addition to the Fairtrade price, to be invested in social, environmental or economic development projects, decided by farmers or plantation workers.

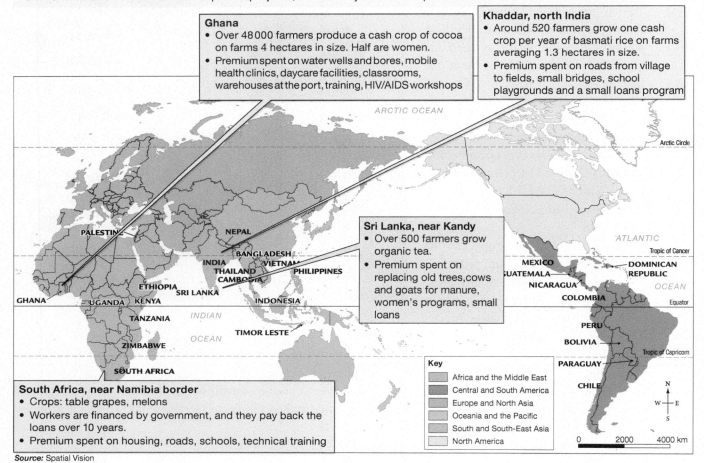

Ghana
- Over 48 000 farmers produce a cash crop of cocoa on farms 4 hectares in size. Half are women.
- Premium spent on water wells and bores, mobile health clinics, daycare facilities, classrooms, warehouses at the port, training, HIV/AIDS workshops

Khaddar, north India
- Around 520 farmers grow one cash crop per year of basmati rice on farms averaging 1.3 hectares in size.
- Premium spent on roads from village to fields, small bridges, school playgrounds and a small loans program

Sri Lanka, near Kandy
- Over 500 farmers grow organic tea.
- Premium spent on replacing old trees, cows and goats for manure, women's programs, small loans

South Africa, near Namibia border
- Crops: table grapes, melons
- Workers are financed by government, and they pay back the loans over 10 years.
- Premium spent on housing, roads, schools, technical training

Key
- Africa and the Middle East
- Central and South America
- Europe and North Asia
- Oceania and the Pacific
- South and South-East Asia
- North America

Source: Spatial Vision

Do farmers benefit from Fairtrade?

Fairtrade farmers enjoy the certainty of having a fixed price and a guaranteed market for their product. This gives them the ability to undertake farming improvements. The Fairtrade Premium provides money for infrastructure, training, education and medical services. It also allows women farmers access to international trade. In Ghana, training of farmers improved the yield by as much as 50 per cent, and it improved school attendance for children by 26 per cent. Fairtrade also benefits the environment, as many of the farming groups produce organic crops.

6.7 Activities

To answer questions online and to receive **immediate feedback** and **sample responses** for every question, go to your learnON title at www.jacplus.com.au. *Note:* Question numbers may vary slightly.

Remember

1. Refer to figure 2.
   (a) Which countries have Fairtrade farmers?
   (b) On which continents are most Fairtrade consumers found?

Explain

2. In Ghana, improving farmer education led to an increase in crop yield of 30 to 50 per cent, as well as improved school attendance. Explain this connection.
3. Read the annotations on figure 2. What similarities are there between Fairtrade farmers in different *places*?
4. What does being an 'active consumer' mean? Why are some willing to pay more for Fairtrade chocolate?

Discover

5. Investigate your favourite chocolate brand. Who owns the brand, and what are their attitudes to Fairtrade chocolate? Write a letter to them expressing your views about Fairtrade cocoa.
6. Research an example of a Fairtrade crop. What are the advantages and disadvantages of Fairtrade?

Predict

7. Predict the future of Fairtrade. What factors might affect your prediction?

Think

8. Undertake a class debate on the topic, 'Fairtrade farmers will improve the future food supply'.
9. Analyse figure 1. What makes it an effective poster? Alternatively, design a poster about one of the Fairtrade crops. Include *change* or *place* or *sustainability* in your poster.

6.8 SkillBuilder: Constructing and describing proportional circles on maps

WHAT ARE PROPORTIONAL CIRCLE MAPS?

Proportional circle maps are maps that incorporate circles, drawn to scale, to represent data for particular places.

Go online to access:

- a clear step-by-step explanation to help you master the skill
- a model of what you are aiming for
- a checklist of key aspects of the skill
- a series of questions to help you apply the skill and to check your understanding.

FIGURE 1 Growth of megacities over time, 1950–2025 (projected)

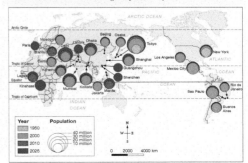

6.9 How do dietary changes affect food supply?

6.9.1 How are diets changing?

According to the Red Cross, 1.5 billion people in the world are dangerously overweight, while 925 million people are underfed. People in both of these categories have health problems. Diets have changed over time and continue to change. This is especially the case in China and India, where the standard of living is rising and people can afford access to a wider variety of foods.

Modern diets have been changing and are expected to change in future (see figure 1). One-third of the world's grain crop is fed to animals to produce meat. Some people consider this wasteful, as it takes about 11 times as much grain to feed a person if it passes through a cow first. While 1500 litres of water is needed to produce 1 kilogram of cereal, 15 000 litres is needed to produce 1 kilogram of meat.

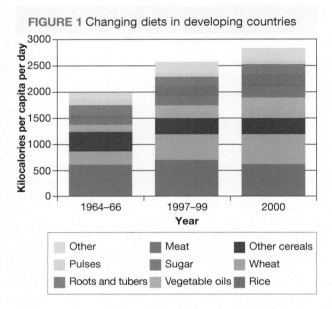

FIGURE 1 Changing diets in developing countries

Legend: Other, Pulses, Roots and tubers, Meat, Sugar, Vegetable oils, Other cereals, Wheat, Rice

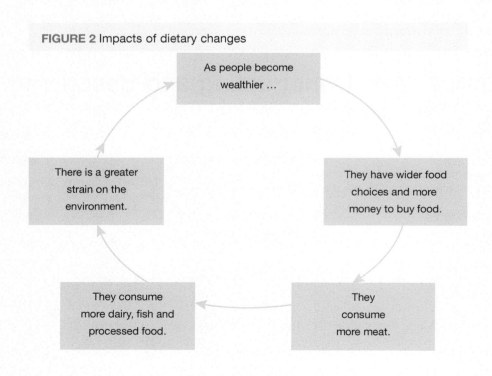

FIGURE 2 Impacts of dietary changes

As people become wealthier …

They have wider food choices and more money to buy food.

They consume more meat.

They consume more dairy, fish and processed food.

There is a greater strain on the environment.

Changing diets in Asia

Rice is a valuable source of protein, but as people's incomes grow, per capita rice consumption is expected to decline. Rice is being substituted by food containing more protein and vitamins and by more processed foods. Japan, Taiwan and the Republic of Korea have already made this dietary change, and the rest of the Asian countries are expected to follow as their economies change and people's incomes increase.

For centuries, the typical Chinese diet was rice and vegetables, occasionally supplemented by small amounts of meat or fish. In 1962, the average Chinese person ate 4 kilograms of meat a year. By 2005, this figure was 60 kilograms and rising (see figure 3). North Americans currently consume the most meat of any country, eating an average of 312 grams per person, per day. This is significantly more than the global average of 113 grams per person.

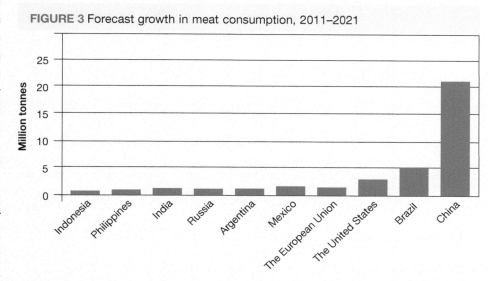

FIGURE 3 Forecast growth in meat consumption, 2011–2021

How can dietary change enable sustainable food production?

Meeting the needs of future populations is not just the responsibility of farmers and producers. We as consumers can also contribute. Attitudes may need to change towards what and how we eat.

- If we are to feed nine billion people sustainably in 2050, it is unlikely to be on a **Western-style diet**, which is rich in meat.
- The world produces enough food to feed 10 billion people. However, a significant portion of our crops is used to feed animals or is used as biofuel to produce energy.
- A switch to a diet containing more plant material would allow land currently used to produce animal feed to instead grow crops to feed humans. Although such a huge change is unlikely, even a small shift can have an impact.
- The Meatless Monday campaign encourages people to go without meat for one day per week. This small change would benefit human health and the health of the planet. Meat production requires a large amount of land, water and energy. Cattle are the largest source of methane gas, which is one of the main contributors to greenhouse gases.
- It is estimated that there are 23 000 edible plants that we do not eat yet. Old foods such as quinoa (pronounced *keen-wah*) could be included in our diets. A crop from South America, quinoa was used over 4000 years ago by the Incas. It has high nutritional value, and grows in a wide variety of climatic conditions. Another advantage of the crop is that all parts of it can be eaten. Peru and Bolivia supply 99 per cent of the world's rapidly expanding quinoa demand. Many countries are investigating its suitability for their locations. The FAO designated 2013 as the International Year of Quinoa.

6.9.2 Can Australia be a food bowl for Asia?

By 2020, it is predicted that half of the world's population will be living in countries in Asia. Four billion people across Asia will experience economic growth of about 10 per cent per year, representing unparalleled opportunities for Australian farmers and the Australian economy. As Asian societies become more affluent, the people are requiring higher standards of living and this includes more and varied foods and a greater quality of fibre. Australian farmers are reported to have clean and green agricultural systems and this, coupled with our location, should provide an economic advantage for our farmed goods.

'We have the potential for a new golden era of Australian agriculture, given the rise of Asia,' our prime minister said in 2012. The challenge for Australian farmers will be in meeting this booming global need for food and fibre by increasing production at a time when we have less arable land, less water and fewer people working in agriculture.

6.9 Activities

To answer questions online and to receive **immediate feedback** and **sample responses** for every question, go to your learnON title at www.jacplus.com.au. *Note:* Question numbers may vary slightly.

Remember
1. Refer to figure 1. What crops are people eating more of? What crops are they eating less of?
2. Refer to figure 2. What is the *connection* between diet and economic development?
3. Refer to figure 3. Describe the *changes* in meat consumption by North Americans and Chinese over time.

Explain
4. Why have people's diets *changed* over time?
5. Explain the *interconnection* between food and your family traditions and celebrations.

Discover
6. How has your family's diet *changed* over time? Ask your parents and family to describe foods and cooking methods from when they were young.

Predict
7. Predict what the consumption of meat in China and the United States might be in 2020. What did you base your prediction on?
8. Predict where your foods might come from in 2050. Could aquaponics or vertical farming be a source of your future food?

Think
9. A United Nations report stated that 'As *changing* the eating habits of the world's population will be difficult and slow to achieve, a long campaign must be envisioned, along with incentives to meat producers and consumers to *change* their production and dietary patterns. Healthy eating is not just important for the individual but for the planet as a whole'. Design a television commercial to promote a Meatless Monday campaign.
10. What might be some of the issues confronting Australia as it attempts to become the 'food bowl of Asia'? What advantages does Australia have in this attempt? How might a farmer react to this suggestion?

 RESOURCES — ONLINE ONLY

 Try out this interactivity: What are we eating? (int-3331)

6.10 Can urban farms feed people?

6.10.1 What are the advantages of urban farming?

In many industrialised countries, it takes over four times more energy to move food from the farm to the plate than is used in the farming practice itself. Properly managed, urban agriculture can turn urban waste (from humans and animals) and urban wastewater into resources, rather than sources of serious pollution.

In 2000, about 15 to 20 per cent of the world's food supply came from urban gardens; in 2015, over 800 million people practised urban agriculture contributing to more than 20 per cent of all global agricultural production.

Farming is usually associated with rural areas, but a growing trend in food production is urban farming. This involves the growing of plants and raising of animals within and around cities, often on unused spaces — even the rooftops of buildings.

Benefits of urban farming include:

- increasing the amount, variety and freshness of vegetables and meat available to people in cities through sustainable production methods
- improving community spirit through community participation, often including disadvantaged people
- incorporating exercise and a better diet into people's lives, leading to improved physical and mental health
- using urban waste water as a resource for irrigation, rather than making it a source of serious pollution
- spending a smaller percentage of people's income on food.

Urban farming could become more important with rapid urbanisation. By 2020 the developing countries in Africa, Asia and Latin America will be home to 75 per cent of all urban dwellers. They will face the problems of providing enough food and disposing of urban waste.

CASE STUDY 1

Container fish farming

On a smaller scale, a German company has developed a sustainable form of aquaculture that can be used in small spaces in cities. It is called **aquaponics**. Fish swim in large tanks in a recycled shipping container (see figure 1). These structures can be located on rooftops and in car parks. The fish waste fertilises tomatoes, salad leaves and herbs growing in a greenhouse mounted above the tank, and the plants purify the water, which is returned to the tanks.

These sustainably produced fresh vegetables and fish can be delivered to nearby city markets and shops, reducing the distance that the products have to travel. The aquaculture containers can be set up almost anywhere, and farmers only need to feed the fish and to keep the fish-tank water topped up. Electric pumps move the ammonia-rich water into the **hydroponic** vegetable garden in the greenhouse.

FIGURE 1 Urban farming – fish and agriculture

Kolkata sewage ponds

The East Kolkata wetlands (see figure 2) cover 12 500 hectares and contain sewage farms, vegetable fields, pig farms, rice paddies and over 300 fishponds. With a population of over 14 million, the Indian city of Kolkata produces huge volumes of sewage daily. The wetlands system treats this sewage, and the nutrients contained in the waste water then sustain the fish ponds and agriculture. About one-third of the city's daily fish supplies come from the wetlands, which are the world's largest system for converting waste into consumable products. The wetlands are also a protected **Ramsar site** for migratory birds. However, the area is now under pressure from urban growth and from the subsequent increase in waste that it needs to treat.

FIGURE 2 The Kolkata wetland system

6.10 Activities

To answer questions online and to receive **immediate feedback** and **sample responses** for every question, go to your learnON title at www.jacplus.com.au. *Note:* Question numbers may vary slightly.

Remember

1. What are the main features of urban farming?
2. What functions do the East Kolkata wetlands perform?
3. How do communities benefit from urban farms?

Explain

4. Suggest what the advantages and disadvantages might be of producing food on the rooftop *spaces* of city buildings. What factors might influence the types of food that could be produced on rooftops?
5. Use the **Aquaponics** weblink in the Resources tab, and outline the advantages of aquaponics presented in the video.

Discover

6. Use the **Vertical farming** weblink in the Resources tab to help you understand vertical farming.
   (a) Draw an annotated diagram to illustrate vertical farming.
   (b) Research an urban farming project in a city. Present it as a PowerPoint presentation.

Predict

7. (a) Predict the *places* in the world likely to have vertical farms.
   (b) Explain why you selected these *places*.
8. (a) Could urban farms encourage agricultural tourism?
   (b) In future, would you consider visiting an urban farm while on holiday?
   (c) Examples of urban farms as agritourism already exist. Can you name any such *places*?

Think

9. Write a letter to the minister for planning, suggesting that urban farming *spaces* should be included in every new urban development.
10. Design a new housing estate with a community garden. What would be needed in order to set up a community garden?
11. When investigating urban farms and people's gardening activities in Denver, United States, researchers found that:
    - people's community pride improved
    - graffiti and vandalism decreased
    - gardeners felt a greater *connection* with their local *place*.

    Are these worthwhile results from urban farming? Explain.

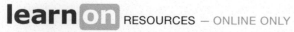 RESOURCES — ONLINE ONLY

🔗 **Explore more with these weblinks:** Aquaponics, Vertical farming

6.11 Review

6.11.1 Review

The Review section contains a range of different questions and activities to help you revise and recall what you have learned, especially prior to a topic test.

6.11.2 Reflect

The Reflect section provides you with an opportunity to apply and extend your learning.

Access this subtopic at **www.jacplus.com.au**

TOPIC 7
Geographical inquiry:
Biomes and food security

7.1 Overview

Numerous **videos** and **interactivities** are embedded just where you need them, at the point of learning, in your learnON title at www.jacplus.com.au. They will help you to learn the content and concepts covered in this topic.

7.1.1 Scenario and your task

Everyone in the world depends completely on the Earth's biomes for the services they provide — from our food and water supply to the regulation of our climate. Over the past 50 years, people have had a more rapid and more extensive impact on these biomes than during any other time in human history. Our demand for food, water, fibres, timber and fuel has driven these changes. The results have contributed to improvements in human wellbeing and economic development, but there has also been detrimental change to many of our major food-producing biomes.

Your task

Your team has been selected to create a website that not only grabs people's attention but also informs them of the importance of one particular biome as a producer of food, and the current threats to food production. Looking into the future, you will also suggest more sustainable ways of managing this biome.

7.2 Process

- Watch the introductory video lesson. You can complete this project individually or invite members of your class to form a group.

7.2.1 Process

You will need to research the characteristics of a biome and address the following four key inquiry questions:

- What is the biome and what are its characteristics and distribution across the world's spaces?
- How can we sustainably feed future populations using this biome?
- In what ways has food production changed this biome? Include examples and/or case studies.
- What are the main types of food production in this biome? How are foods produced?

Each group should decide how to divide the workload so that each of the four inquiry questions is studied.

7.2.2 Collecting and recording your information and data

- Once you have chosen your biome and divided the key questions among the team, it is time to start researching information. For your own key question, break it down into several minor questions that can become subheadings to form the structure of your research. As a group, check each person's research structure to ensure that it follows the inquiry sequence.

When researching, look for maps, graphs and images that support your key question or that of another team member. You should also look for data or statistics that you can show visually in the form of maps, diagrams or graphics.

7.2.3 Analysing your information and data

- Once you have researched and collected relevant information, you need to review it, ensure that you understand the material and then use it to answer your key questions. From maps and graphs, describe any patterns or trends that you identify. If using photographs, write clear annotations for each one, highlighting particular features.
- Access the website model and website-planning template from the Resources tab to help you build your website as well as images and audio files to help bring your site to life.
- Use the website-planning template to create design specifications for your site. You should have a home page and at least three link pages per topic. You might want to insert features such as 'Amazing

facts' and 'Did you know?' into your interactive website. Remember the three-click rule in web design — you should be able to get anywhere in a website (including back to the homepage) with a maximum of three clicks.

7.2.4 Communicating your findings

- You will need to access FrontPage, Dreamweaver, iWeb or other website building software to build your website. Remember that less is more with website design. Your mission is to engage and inform people about a topic they may never have thought about. You want people to take the time to read your entire website.

7.3 Review

7.3.1 Reflecting on your work

- Think back over how well you worked with your partner or group on the various tasks for this inquiry. Reflect on your contribution to the team by completing the Reflection template in the Resources tab. Determine strengths and weaknesses and recommend changes you would make if you were to repeat the exercise. Identify one area where you were pleased with your performance, and an area where you would like to improve. Write two sentences outlining how you might be able to do this.

- Print out your Research Report from the Resources tab and hand it in with your website and reflection notes.

UNIT 2
GEOGRAPHIES OF INTERCONNECTIONS

Every text, call, purchase or trip we make connects us to information, other people and places. This interconnection is influenced by people's views or perceptions of these places. Our consumption of goods and services and our travel, recreational and cultural choices all have impacts on the environment. This has implications for future sustainability.

8 How do we connect with places?   135

9 Tourists on the move   158

10 Buy, swap, sell and give   183

11 For better or worse?   208

12 Fieldwork inquiry: What are the effects of travel in the local community?   232

Planet Earth at night, lit by the rising sun, illuminated by the light of cities. The surrounding network represents the major global air routes.

TOPIC 8
How do we connect with places?

8.1 Overview

Numerous **videos** and **interactivities** are embedded just where you need them, at the point of learning, in your learnON title at www.jacplus.com.au. They will help you to learn the content and concepts covered in this topic.

8.1.1 Introduction

Geography is the study of people and their connections with places. The way we interact with places is dynamic: we change places and places change us. In a world of over seven billion people, we have many different perceptions of what a place is like, how it is used and how it could be improved. More people are on the move, too. Their journeys may be on foot or by plane as they visit and interact with new places. With rapid developments in ICT, some of those places may be imagined. What do our connections look like today, and how will they change tomorrow?

An Elder explains the meaning of an indigenous rock art cave painting in Cooktown.

Starter questions

1. How do we as individuals see or perceive *places*?
2. How do our perceptions differ from those of our friends, our neighbours or people on the other side of the world?
3. How do distance and time play a role in our perception of *places*?
4. Do we make assumptions about *places* and what they are like?
5. How do *places change* us?
6. Is ICT *changing* the way we look at *places*?

INQUIRY SEQUENCE

8.1 Overview		135
8.2 How do we 'see' places?		136
8.3 **SkillBuilder:** Interpreting topological maps	online *Only*	138
8.4 How do we move around our spaces?	online *Only*	139
8.5 What does our land mean to us?		139
8.6 How do places change?		142
8.7 How do we access places?		144
8.8 To walk or not to walk?		149
8.9 **SkillBuilder:** Constructing and describing isoline maps	online *Only*	153
8.10 Are we all on an equal footing?		153
8.11 How do we connect with the world?	online *Only*	157
8.12 **Review**	online *Only*	157

8.2 How do we 'see' places?

8.2.1 Perceptions of places

People's **perceptions** of places are rarely the same. A person's view of a particular place or region is coloured by their own culture, experiences and values. The characteristics and significance of a place will be viewed differently by each individual, and our mental maps of the world can change daily as we have new experiences and gain new knowledge.

The biggest influences on the way we perceive places are age, gender, class, language, **ethnicity**, race, religion and values. How important a place is to us may be determined by whether we feel that place belongs to us or not, whether it is part of our tradition or history, or whether the place is totally unfamiliar.

A place can seem exciting, scary, interesting or boring depending on our experience, expectations or mood on a particular day. Our perceptions of places may also change over time according to climatic changes, conflict or economic shifts.

It is important to understand the factors that influence our perceptions of places and regions, as well as the impact that other groups and cultures have on our perceptions. If we can understand those influences, we may be able to avoid the dangers of **stereotypes** and appreciate the diversity that exists around us.

FIGURE 1 The Kaaba in Mecca — sacred, interesting or crowded?

8.2.2 How do we map places?

We all form an impression of our physical surroundings — even of places we have never actually been to. These are what geographers call our mental maps.

Mental maps tell us how we order the space around us. There is no such thing as an accurate mental map, but people's mental maps of their immediate environment tend to be more realistic than those of places they have never visited. Think about some of the ways you use mental maps in your daily life. You may direct someone from point A to point B, telling them about landmarks they will see along the way. You may think about the quickest way to get to the city from a friend's house, imagining your route in your mind.

Our mental maps can help document our influences. Those who walk a lot may be more connected with their neighbourhood and surrounding environment, whereas those who drive will have a very different perspective in their mental map. In the 'Streets Ahead' study by VicHealth, children who walked to school drew pictures that included street names and friends' houses, and they were able to describe people and places in detail (see figure 2a). Children who were driven to school tended to separate items from their environment, displaying them in distinct windows (see figure 2b).

Mental maps of places we are unfamiliar with are heavily influenced by the media and stereotypical discussions. Travel helps to counteract the effects of the media and generally increases a person's knowledge of an area, providing them with a better understanding of what a place is really like.

FIGURE 2 (a) Drawing by a child who walks to school (b) Drawing by a child who is driven to school

(a) **(b)**

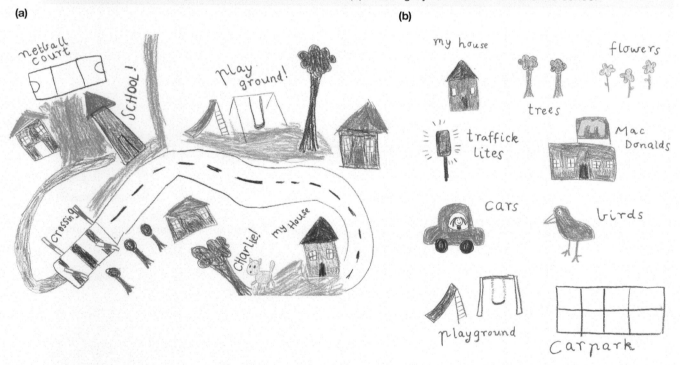

8.2 Activities

To answer questions online and to receive **immediate feedback** and **sample responses** for every question, go to your learnON title at www.jacplus.com.au. *Note*: Question numbers may vary slightly.

Remember

1. List the factors that may influence our perception of *place*.

Explain

2. From your list in question 1, which factor do you think is most influential? Why?
3. Why do you think the two children's maps shown in figure 2 are so different? What does it say about their *interconnection* with their *environment*?

Discover

4. With your class, make a list of the *places* or landmarks in your community that you use on a regular basis. Each student should rate the importance of each on a *scale* of 1 to 3, with 3 being most important. Collate the data to find out which *places* are most and least important to your class. Are the results as you expected? Do they match your own perceptions of how important *places* are, or do you have a different view from your classmates? Explain why there might be similarities or differences.
5. Create a mental map of your journey to school on a blank sheet of A3 or A4 paper. Include as many annotations as you can, such as street names, landmarks, shop names and so on. Once you have finished, compare and contrast the *scale*, size and accuracy of your mental map with a street directory or an ICT mapping tool. Write a paragraph that details some of the differences between your perception and reality.

Think

6. How would the *place* in figure 1 be viewed by different groups? What kind of experiences or influences may affect their view? Try to provide at least three perceptions for the image.
7. The writer Henry Miller once said, 'One's destination is never a *place*, but a new way of seeing things'. What does this quote mean to you, in light of your knowledge of various *places*, your own travels, and what you have learned about perception?

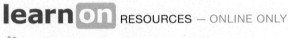

learn on RESOURCES — ONLINE ONLY

🔧 **Try out this interactivity:** My place (int-3332)

8.3 SkillBuilder: Interpreting topological maps

on line only

WHAT IS A TOPOLOGICAL MAP?

Topological maps are very simple maps, with only the most vital information included. These maps generally use pictures to identify places, are not drawn to scale and give no sense of distance. However, everything is correct in its interconnection to other points.

Go online to access:

- a clear step-by-step explanation to help you master the skill
- a model of what you are aiming for
- a checklist of key aspects of the skill
- a series of questions to help you apply the skill and to check your understanding.

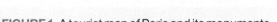
FIGURE 1 A tourist map of Paris and its monuments

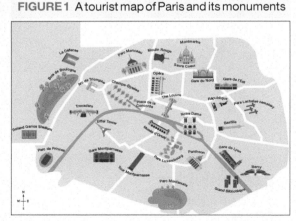

learn on RESOURCES — ONLINE ONLY

🎞 **Watch this eLesson:** Interpreting topological maps (eles-1736)

🔧 **Try out this interactivity:** Interpreting topological maps (int-3354)

8.4 How do we move around our spaces?

Access this subtopic at **www.jacplus.com.au**

8.5 What does our land mean to us?

8.5.1 Why is land so important?

Land means many things to different people. A farmer sees land as a means of production and a source of income. A conservationist sees land as a priceless natural resource that must be protected. A property developer sees land as an area that can be divided, built upon and sold for a profit. To Indigenous Australian peoples, land is something much, much more — it is a part of their being.

Indigenous Australian peoples have been in Australia since the beginning of the Dreaming (more than 50 000 years by European estimates), adapting to survive and thrive in a changing environment. For Indigenous Australian peoples, the land is at the core of their wellbeing — their relationship with the land is one of interconnectedness across the physical, spiritual and cultural worlds (see figure 1).

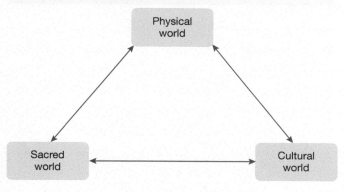

FIGURE 1 A simplified view of the Indigenous Australian peoples relationship with the land

8.5.2 How do Indigenous Australian peoples perceive the land?

For Indigenous Australian peoples, land is much more than the soil, rocks, hills and trees. The land is where they come from, and to where they will return. The land, or country, represents a whole environment that sustains Indigenous Australian peoples and their culture and way of life. Indigenous Australian peoples are diverse, made up of over 500 different groups, each with its own separate language (or dialect), laws, beliefs and customs. Language groups are made up of a number of communities, with each community belonging to a territory or traditional land. These places include features of the natural environment such as waterholes and hills, as well as distinct geographical boundaries such as rivers or mountain ranges. Natural features are often represented in Indigenous art (see figure 2).

It is the responsibility of each community to look after their country as it looked after them. The environment holds rich meaning for Indigenous Australian peoples, whose Dreaming Stories (for Aboriginal peoples) and Legends (for Torres Strait Islander peoples) are present throughout the landscape, along with many sacred places for special ceremonies — men's and women's sacred sites (see figure 3) — and resting places for ancestors that must be protected and conserved.

Each community has a **totem** that was a sign of its people's spiritual link to the land. A totem could be an animal, plant or geographical feature such as a weather pattern or rock formation. It is from this totem or land feature that an individual draws their spirituality, and they would feel a special responsibility to protect it. Special ceremonies are performed at these sacred places to show respect for, replenish and celebrate each totem.

8.5.3 Does the land belong to us or do we belong to the land?

Indigenous Australian peoples view themselves as custodians rather than owners of land, for the land will exist long after they have left this world. To them, land cannot be bought or sold. The concept of property or land ownership that arrived with the Europeans contrasted greatly with the indigenous view of place.

FIGURE 2 A traditional Aboriginal dot painting depicting land at Kiwirrkura, 400 kilometres west of Alice Springs

 Water, rainbow, snake, lightning, string, cliff or honey store

 Waterholes connected by running water

 Camp site, stone, waterhole, rock hole, breast, fire, hole or fruit

Source: © Donkeyman Lee Tjupurrula Kukatja (c.1921)–1994

Tingari Dreaming at Walawala (1989)
Synthetic polymer paint on canvas, 119.7 × 179.3 cm
Purchased from Admission Funds, 1989
National Gallery of Victoria, Melbourne
© Donkeyman Lee Tjupurrula/Licensed by VISCOPY 2013

When the European colonies were established, many Indigenous Australian peoples were dispossessed of their land and cultural practices were forcibly disrupted. In many cases, indigenous communities were pushed onto marginal lands that were often not their own, not only creating conflict but severing their connection with the land from which they drew their sense of identity. However, even today among groups largely displaced from their traditional estates, that strong link to country is maintained through stories and a sense of place and spiritual connection.

8.5.4 How do you view this land?

To illustrate this difference in viewpoint, figure 3 shows James Price Point on Western Australia's Kimberley coast. The following are three very different views of the same area of land.

• Unremarkable beach — *Colin Barnett, Premier, Western Australia*
• Major heritage site — *WA Department of Aboriginal Sites* ('Major' is the Department's highest category.)
• Secret Aboriginal men's business site — *Goolarabooloo Aboriginal people*

When different people have vastly different views about a place, it can make the management of that land challenging.

FIGURE 3 James Price Point, Western Australia

8.5.5 How did Indigenous Australian peoples move across the land?

Much like the international system of passports and visas to enter other countries, a similar process exists for Indigenous nations. Entry to another nation's or community's lands is by ceremony and negotiation, a practice still commonplace today, recognising the important relationship that Indigenous Australian peoples have with their country. The tradition of a 'Welcome to Country' for visitors issues a shared commitment to protect and preserve the land being visited. After being welcomed, those who walk on another's lands are expected to respect the traditional owners' rules and protocols.

8.5.6 What places do Indigenous Australian peoples connect with today?

In the twenty-first century, 78.6 per cent of Aboriginal and Torres Strait Islander peoples are interconnected with Australia's urban environments; only 21.4 per cent live in remote or very remote areas. Many of those living in urban environments know the stories passed through generations, but not all have visited their traditional lands. AFL footballer and Australian of the Year 2014, Adam Goodes took time out from his career to return to his homeland, to find out more about himself and his people, and to help identify characteristics that made him different.

8.5 Activities

To answer questions online and to receive **immediate feedback** and **sample responses** for every question, go to your learnON title at www.jacplus.com.au. *Note*: Question numbers may vary slightly.

Remember

1. How do Indigenous Australian peoples perceive the land?

Explain

2. What does land mean to you? Think about where you live or where you come from to help describe the *interconnection* you have with the land.
3. Brainstorm with other members of your class and construct a list of other examples of different cultural viewpoints to the same object, custom or *place*. Consider such things as music, religious customs and foods.

Discover

4. Refer to figure 3 and conduct internet research to find out more about James Price Point. Answer the following questions.
   (a) What is involved in the conflict over James Price Point?
   (b) Create a mind map that shows the various groups involved in the dispute. Beneath each group name, list their interests in the site.

(c) Consider the viewpoints about James Price Point quoted in this section. How have these individuals or groups perceived the land in this *place*?

(d) The project at James Price Point has recently been cancelled. From your internet research, why do you think this happened? Share your findings with your classmates. Do they agree?

5. Who are the traditional owners of the land on which you live? Have you witnessed a 'Welcome to Country' ceremony? Who performed the ceremony, and what was involved? (It may have included a speech, traditional dance or smoking ceremony.)

Think

6. Given the strong *interconnection* to land, you may think that Indigenous Australian peoples are opposed to land development. Although custodial responsibilities and care of the land are of utmost importance, many land owners strongly support economic development. As a class or in small groups, debate the arguments in favour of and against development of traditional lands.

8.6 How do places change?

8.6.1 A chequered history

Places can change very slowly over time and space, or undergo rapid transformations. Melbourne's laneways are an excellent example of how a place once perceived as unsanitary and unsafe is now a thriving and popular part of a metropolis.

During the gold rushes of the 1850s, Melbourne's laneways were well used by people from all walks of life. Then, at the turn of the twentieth century, they began to take a turn for the worse. Criss-crossing the city, their main function was as a place for rubbish disposal. They were dark and dingy, and riddled with disease, crime, gambling houses and brothels. After two World Wars, they became home to many immigrants who had nowhere else to live. The city had lost its shine.

Then, in the late twentieth century, something changed. Perhaps influenced by the potential for regeneration that they had seen in European cities on their travels, people began to see the potential of Melbourne's neglected laneways. Small businesses such as art and craft galleries, fashion boutiques and music shops opened. Business owners leased cheap properties in the laneways, away from the main streets with their high rents. Public spaces were regenerated, adding to the city's landscape. Music and entertainment became a reason to go into the city at night.

People are living in the city again, and the CBD is now perceived as a desirable address — its resident population is now approximately 30 000 compared with only 700 in the 1980s. The laneways have helped lead that revival.

The laneways today

Better lighting, more cleaning and an increased number of people have all contributed to a change in the perception of Melbourne's laneways. **Street art** tours abound, and many laneway bars, cafés and restaurants are desirable places to see and be seen in. There are laneway festivals, parties and even a 'Love your Laneway' project run by the City of Melbourne. The laneways are one of Melbourne's biggest tourist drawcards, and are particularly famous for the vibrant and colourful street art that adorns their walls. Rather than simple **graffiti** or tagging, these are inspiring artworks from some of Australia's (and occasionally the world's) best street artists.

8.6.2. The laneway revival around Australia

The city of Brisbane has also undergone a transformation in the last few years. The Brisbane City Council's Vibrant Laneways initiative, combined with the introduction of the state government's small bar licence, has prompted many innovative venues to open their doors. In 2012, Adelaide also announced reduced liquor licensing fees for smaller venues. This initiative is aimed at revitalising the Adelaide CBD, creating a local version of laneway bars and culture akin to Melbourne's CBD. In New South Wales, the City of Sydney's Laneway Art Program has sought to encourage a city that buzzes with new art, new business and new life.

FIGURE 1 A map of Melbourne's laneways (a) Centre place, one of Melbourne's revitalised laneways
(b) Duckboard place (c) ACDC lane (d) Graffiti in Hosier Lane

(a)

(b)

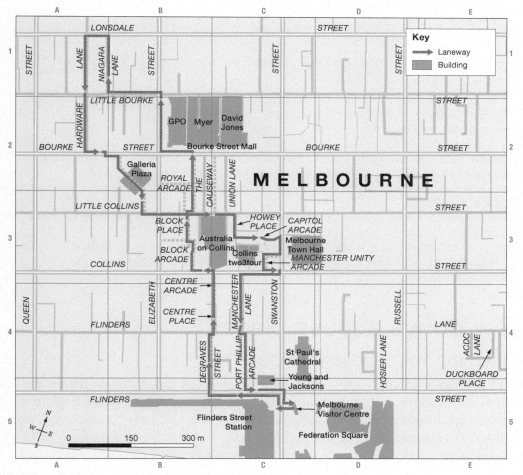

Source: Spatial Vision

(c)

(d)

8.6 Activities

To answer questions online and to receive **immediate feedback** and **sample responses** for every question, go to your learnON title at www.jacplus.com.au. *Note*: Question numbers may vary slightly.

Explain

1. Explain how the perception and uses of Melbourne's laneways have *changed* over time.
2. How do laneways allow people to *interconnect* with the city?

Discover

3. What other areas in Melbourne, or your city, are now developing a laneway culture? How are they attempting to achieve this? In Melbourne, you may wish to investigate Richmond, the QV building or the Docklands precinct.

Predict

4. What other uses could you propose for laneway *spaces* in addition to those outlined above?

Think

5. Some aspects of laneways that can be improved are:
   • waste management and stormwater run-off
   • amenity and access
   • infrastructure, such as public lighting and road surfaces.
   (a) Can you think of any other ways in which laneways can be improved for public use?
   (b) Are there other *spaces* within a CBD *environment* that could be improved in order to provide new *places* for people to enjoy?
   (c) What would need to happen in order to make the *places* you identified in question 5b functional, safe and accessible?
6. If possible, conduct fieldwork in your city's laneways or complete the Laneway Walk shown in the map in figure 1. Are laneways *sustainable spaces* for people? Give a detailed personal response.

8.7 How do we access places?

8.7.1 Connecting with public transport

Public transport provides a relatively low-cost way for people to interconnect with places, and can reduce traffic congestion and pollution. For students, it is often the only way to get around. Sometimes, however, it can seem like too much bother, perhaps because one service does not connect to another or because there are not enough services running, especially near your house.

Public transport use is considerably higher in capital cities than in other parts of Australia, partly because cities have relatively large populations and better public transport **infrastructure**.

8.7.2 Our changing needs

With any population growth, governments at all levels must consider how they will meet changing transport needs. ICT developments have allowed us to make better decisions for our use of public transport. For example, have you used the internet or an app to find the fastest way to get from A to B? Service quality, frequency and infrastructure are generally the biggest concerns in the provision of a public transport system. However, the affordability of public transport is equally important, because many low-income people depend upon public transport to access jobs, services, education and recreation.

8.7.3 Different forms of public transport have different uses

Public Transport Victoria aims to deliver quality customer service and provide enhanced access across the range of public transport services throughout Melbourne and Victoria via its website, apps and other forms of social media.

Trains move large numbers of people over long distances at high speed in and out of the Central Business District. Travellers' access is heightened by the use of curving routes across the urban area with a minimum number of stops.

Trams operate only in areas of high population density, using high-speed and infrequent stops to maximise access for middle-distance commuters.

Buses provide access where trains and trams do not go and 'infill' access for people by using a range of road levels. Buses are the most flexible of the services, able to change routes as there is no fixed rail system involved. Buses, and to some extent trams, ferry people to train stations, adjusting timetables and reorganising routes to match the train network, as seen with the opening of the Regional Rail Link in Melbourne's outer west.

FIGURE 1 Six trains come and go from central Melbourne.

FIGURE 2 Trams take people from the city to suburbs such as Maribyrnong in the north west.

FIGURE 3 Buses and trains interconnect at Glen Waverley station.

8.7.4 How satisfied are we with public transport?

The Transport Opinion Survey is conducted biannually across 1000 adult Australians. In 2015, 37 per cent of those surveyed said public transport was a top priority. Three in five Australians would prefer to see rail developments rather than more bus services, except in Queensland and South Australia, which already have well developed, fast bus systems. Only 9 per cent of those surveyed thought Australia's transport systems would be better within a year; just 7 per cent felt that local public transport would be improved within a year. In five years' time, 22 per cent of those surveyed thought the transport system they were using would be better.

FIGURE 4 Shared pathways are common.

A recent study in the Netherlands revealed that people perceive that their travel time on public transport takes 2.3 times as long as driving a car to make the same journey. People also perceive a continuous journey (involving, say, only one train) as taking less time than a journey that involves transfers and waiting times, even if the second journey is actually shorter. People estimate the waiting time to be about two to three times longer than the actual time. So, a wait of 10 minutes is perceived as 20 to 30 minutes. Factors that influence this perception include:

- uncertainty about when the next bus or train will arrive
- weather conditions
- familiarity with the journey.

Given that travellers tend to consider non-vehicle travel time (walking, waiting, transferring) to be more difficult than in-vehicle travel time, this has consequences when trying to attract people to public transport. If people think their travel time by car is 60 minutes, they perceive their travel time by public transport for that same trip to be almost double: 117 minutes.

8.7.5 Active travel

Cycling and walking to get to work, to visit friends and for the purpose of recreation have become mainstream modes of transport in the twenty-first century. In particular, in Melbourne's inner suburbs 20 per cent of those going to work now choose to use active travel.

Melbourne's bicycle paths and trails continue to grow in number providing increased access to places. Figure 5 maps the 'spiderweb' of pathways around the Melbourne region.

To encourage active travel, railway stations — both new and old — are installing secure bicycle storage areas (called parkiteers, see figure 6). Authorities are revising road layouts and regulations to provide a secure riding environment. Manufacturers are also designing electric bicycles to make access available to a wider range of people. Bicycles for hire, or free, can be picked up at points within the Central Business District of many capital cities.

The choice to access places by public transport, active travel or vehicle keeps people connected and strengthens interconnection in a community.

FIGURE 5 Melbourne's bike paths

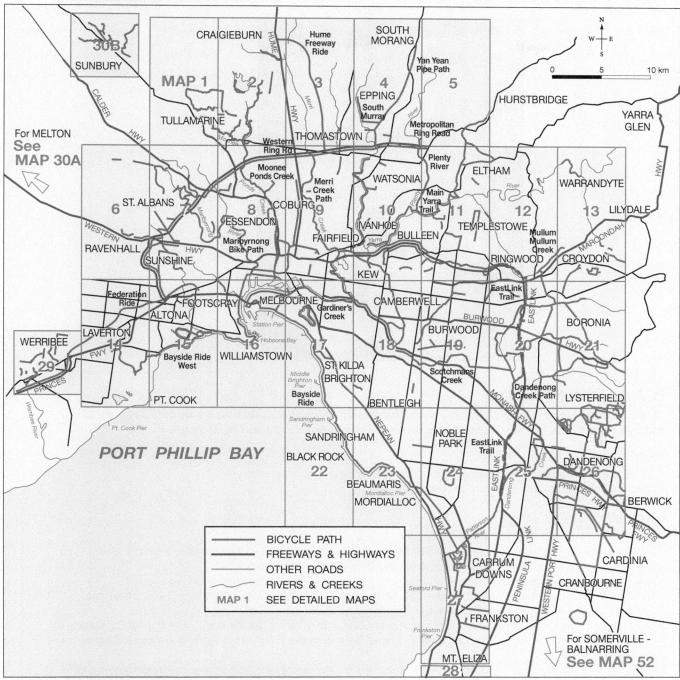

Source: Map courtesy the Bike Paths and Rail trail Guide (Victoria) http://www.bikepaths.com.au

FIGURE 6 Signage at a railway station encouraging the use of parkiteers

8.7 Activities

To answer questions online and to receive **immediate feedback** and **sample responses** for every question, go to your learnON title at www.jacplus.com.au. *Note*: Question numbers may vary slightly.

Remember

1. Why is public transport perceived by governments as being very important?
2. Write a definition for the term 'active travel'.

Explain

3. Do you use public transport? Why or why not?
4. (a) How interconnected is the place in which you live? What types of public transport are available to you? What distances do you need to travel to reach a bus stop or a train station?
   (b) What types of public transport are required for you to access your closest international airport?
   (c) How do you perceive the quality of your public transport? Consider accessibility, timeliness, cleanliness, comfort, ticketing, safety, convenience and information about the service. Explain you answer.
5. Explain how each form of public transport provides access for different groups in our community. Consider students, workers, senior citizens, those with a disability and tourists. Figures 1, 2 and 3 will provide some additional ideas.
6. One of the most significant aspects of public transport is the interconnection between the different forms of transport. Why is interconnection important?
7. See the **Melbourne bike paths** weblink in the Resources tab for a satellite image of Melbourne's bike paths. Go online and create figure 5. Use the scale tool to measure three distances between places that you are familiar with and three distances between places that you would like to visit. How do you perceive your ability to get to these places by bicycle?

Discover

8. (a) As a class, pick a location on the other side of town. Using your rail, bus or other public transport provider website, find out how long it would take you to travel from your school or home to this point on:
   • a Monday morning at 9 am
   • a Sunday evening at 6 pm.

(b) What did you notice about the travel times? Were they different? Why do you think this is? Create a map of your journey, using an appropriate key, to show rail, bus and other modes of transport used.

9. Using the interactive **TRAVIC (Transit Visualization Client)** weblink in the Resources tab, look at the movement on the public transport system of Melbourne and in regional Victoria in real time. How does this impact on your perception of the role of public transport? If possible, visit this site at different times of the day and night to see how the transport system adjusts to the needs of the people.

10. How safe is it to ride a bicycle in Melbourne? Visit the survey on the internet at BikeSpot to find evidence for your opinion.

Think

11. There are frequent announcements by the Victorian Government on developments with the public transport system. For each public transport type and for each form of active travel, suggest two changes that might occur in the next 10 years. Compare your suggestions with those of others in the class.

8.8 To walk or not to walk?

8.8.1 Introduction

Urban planners around the world are focusing on human wellbeing as a key to the structure of new suburbs and revitalisation of existing suburbs. People's perceptions of what will make 'life good' and what makes a 'good place' to live in are being taken into account. Being connected to other places and people is a high priority.

8.8.2 What is the '20-minute neighbourhood'?

Melbourne has been ranked the best city to live in since 2011. The city is unique, with access to coastal areas, a mild climate, a range of topography, distinctive suburbs or places, considerable tree cover and well-designed buildings and streets. As part of 'Plan Melbourne 2016', which aims to help retain that status and enhance the state's capital, the concept of the '20-minute neighbourhood' is being applied.

As figure 1 shows, the '20-minute neighbourhood' is about improving the liveability of a place. This means being able to walk around your neighbourhood and within 20 minutes being able to access your daily needs — for example transport, doctors, primary schools. Factors that make a good neighbourhood walkable are:

- a centre — either as a street or public space
- people — enough people for businesses to be successful and for public transport frequency
- mixed income and mixed use — a range of housing types
- parks and public space — for people to gather and to play
- pedestrian design — foot access (cars parked off street)
- schools and workplaces — close enough to walk
- complete streets — suited to bicycles and walking, and allowing easy movement across the place.

8.8.3 Why is walkability important?

Walkability provides a range of benefits to any community. People's health has been shown to improve if they walk on a regular basis. In particular, the risk of heart disease and diabetes is reduced. When people walk regularly, they are often 2.5 to 4.5 kilograms lighter than they would otherwise be. There is a reduced

FIGURE 1 The components of the '20-minute neighbourhood' concept

environmental impact with fewer cars on the road, as feet produce zero per cent carbon dioxide emissions! Communities benefit when people have more time available for involvement in community activities. Up to 10 per cent of a person's time spent in a community activity is lost when a car is used for just 10 minutes of commuting. Families also benefit financially as a car is often the second largest household expense and housing prices can increase by 20 per cent when located in places with a high walkability score.

In Eugene, Oregon, in the United States, areas closest to the centre of the city were shown to have a higher walkability rating than those on the rural urban fringe. Key elements considered in the mapping and overlaying (figure 2) were:

- density — the density of the population and the number of employees
- destinations — bus stops, shops, primary schools, corner stores, parks and other goods and services
- distance — intersection density, bicycle facilities, paths

FIGURE 2 Eugene, Oregon walkability ratings with the most walkable areas shown in red

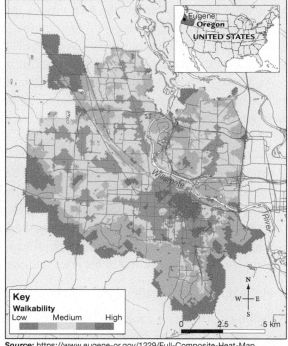

Source: https://www.eugene-or.gov/1229/Full-Composite-Heat-Map

- aesthetics — tree cover, road width, condition of properties along routes
- safety and perceived safety — traffic speed, path condition, signalled crossings
- socioeconomics such as distribution of income, education, age and background.

8.8.4 How accessible are Melbourne's neighbourhoods?

A US company developed a Walkability Index (table 1), which considers a range of features using the Eugene, Oregon experience. According to its scale, the following key assesses the rating of a neighbourhood.

TABLE 1 The classifications within the Walkability Index

Walkability rating	Access to the neighbourhood	Tasks able to be completed
90–100	Walker's paradise	Daily errands do not require a car
70–89	Very walkable	Most errands done by foot
50–69	Somewhat walkable	Some errands done by foot
25–49	Car dependent	Most errands require a car
0–24	Car dependent	Almost all errands require a car

Using this Walkability Index there are just 12 suburbs of Melbourne that currently fit the range of 90–100 as a walker's paradise. These suburbs include Carlton, Fitzroy, Fitzroy North, Melbourne (central area), St Kilda, South Yarra, East Melbourne, South Melbourne, Collingwood, Windsor, Southbank and Richmond.

8.8.5 How accessible will neighbourhoods become?

Property developers across all major cities in Australia and the developed world have realised the importance of human wellbeing. New estates now focus on providing parklands (often with a water feature); local shopping centres; safe and sound surroundings; foot and bicycle paths; and peaceful, clean, green environments. Advertising for these estates centres around building communities, with young families living an active lifestyle.

FIGURE 3 Box Hill is an important transport hub for trains and buses.

FIGURE 4 Development of the Point Cook estate.

Source: Central Equity, Featherbrook Point Cook.

Planners and developers in established suburbs are seeking to 'infill' the suburbs, creating and recreating to form 'the 20-minute neighbourhood' around activity hubs and avoid further encroachment on farming land for urban development. The challenge is to have about a 70 per cent increase in housing and population living in the local area hubs. Community and transport infrastructure will need to be revised to achieve this target. In Melbourne, activity hub development can be found at Box Hill, Broadmeadows, Dandenong, Epping, Footscray, Fountain Gate/Narre Warren, Frankston, Ringwood and Sunshine.

8.8 Activities

To answer questions online and to receive **immediate feedback** and **sample responses** for every question, go to your learnON title at www.jacplus.com.au. *Note:* Question numbers may vary slightly.

Remember

1. Draw a mind map (see section 8.2) of the distances you have to travel from your home to the bus stop or train station, to school, to the shopping centre, to the park where you meet your friends, to a place for sporting activities, and to any other significant locations in your life. Discuss in class how teenagers perceive the distances travelled.
2. Recall and list the features of 'the 20-minute neighbourhood'.

Explain

3. Using figure 2, describe the distribution of the different levels of walkability in Eugene, Oregon.
4. Suggest factors that may influence the location of the high level of walkability in Eugene.
5. To avoid expansion of Eugene:
   (a) Suggest two areas of the city that the city planners and developers might be looking at to improve the level of access. Provide reasons for your choice.
   (b) Suggest a change that may be able to be implemented in the short term, medium term and long term within the city to improve access.

Discover

6. On a map of Melbourne, find the 12 suburbs with a high walkability rating. Describe the locations of these places.
7. Using the internet for 'How Walk Score Works' find:
   (a) the walkability rating for Australia's major cities. Comment on their scores.
   (b) the walkability rating for your place/home. Can you explain why your place has been given its rating?
   (c) the walkability rating of a rural environment that you know. Explain why rural areas might be more car dependent.

Think

8. Many parents don't allow their children to walk to school any more. Make a list of the safety issues that parents perceive about access to school.

Predict

9. In a small group, draw a plan for a 20-minute neighbourhood that you would like to access and live in. Discuss and consider each group member's perception of which features make for wellbeing in a community.

8.9 SkillBuilder: Constructing and describing isoline maps

WHAT IS AN ISOLINE MAP?

An isoline map shows lines that join all the places with the same value. Isoline maps show gradual change in one type of data over a continuous area. Isolines show change in the trend of the data being mapped. Isoline maps are easy to understand, especially when coloured between the lines.

Go online to access:

- a clear step-by-step explanation to help you master the skill
- a model of what you are aiming for
- a checklist of key aspects of the skill
- a series of questions to help you apply the skill and to check your understanding.

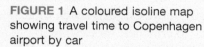

FIGURE 1 A coloured isoline map showing travel time to Copenhagen airport by car

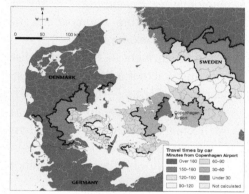

8.10 Are we all on an equal footing?

8.10.1 How accessible are our cities?

Many of us take it for granted that we can walk to the shops, hop on a bus and go to the city centre, or find out when the next train is departing. Accessibility should mean that people with **disabilities** have the same access to the physical environment, transportation, information and communication technologies, and other facilities and services. Everyone should feel connected with society, rather than separated from it.

Our cities can be a depressing obstacle course for millions of people. For those with a disability, negotiating a flight of stairs, opening a door or even reaching a lift button can be impossible. Have you ever considered how difficult our cities can be for some of their citizens and visitors?

8.10.2 Providing equal access

Almost one in five Australians (18.5 per cent) reported a disability in 2012. Four in 10 Australians over the age of 18 report having a disability or long-term health condition. Equal access, particularly to transport, is essential for equality. Limiting transport can mean limiting people's opportunities.

A disability can take many forms, including:

- walking disabilities — cannot use stairs easily, moves slowly, needs wider spaces (due to crutches, for example)
- manipulatory disabilities — has difficulty in operating handles and ticket machines, for example
- vision impairment — has trouble distinguishing between road and pavement, identifying platform edges, knowing whether a lift has arrived at the correct floor, seeing signs or directions
- hearing problems — has difficulty hearing announcements about delays, cancellations or emergencies, or hearing an approaching vehicle
- intellectual disabilities — is challenged by being in an unfamiliar setting, or coping with cancellations or complex timetables
- psychiatric disabilities — experiences stress, anxiety or confusion in crowded situations or encounters with other travellers
- wheelchair disabilities — difficulty moving about when no ramps are available, when there are insufficient or badly designed parking spaces, or when there is not enough room to manoeuvre equipment.

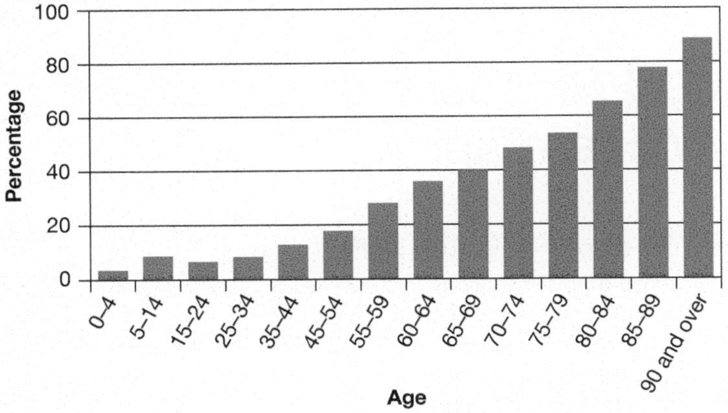

There are also additional (and sometimes less obvious) disabilities to consider, such as asthma, epilepsy, obesity and diabetes, and temporary disabilities that result from injuries that are rarely classified as disabilities. When considering transport disadvantage, we must also include elderly people, low-income earners, children and outer-urban dwellers, who experience this to some degree as well. Parents with prams or strollers may also be affected.

FIGURE 1 Percentage of Australians with a reported disability

8.10.3 A city for everyone

In 2015, the Swedish city of Borås won the European Union's Access City award — a prize for the European city that is most accessible for people with disabilities. The annual honour aims to award efforts to improve accessibility in the urban environment and to foster equal participation of people with disabilities. People with disabilities and their representative organisations were heavily involved in the planning process. Some of the improvements are that:

- a database has been established of all public buildings with adequate access
- route markers have been laid both indoors and outdoors
- public transport is free on easily accessible buses
- digital 'locks without keys' have been provided for easy access, especially for support services.

All these measures benefit the entire community as well as the economy, because when everyone has access to places of work, people feel included and government social support systems are not overloaded.

8.10.4 Accessing cities

The city of Melbourne is keen to have an inclusive environment for all. Its 'Melbourne for All People Strategy Street 2014–2017' has six key points:

- connection
- access and inclusion
- safety
- health and wellbeing
- lifelong learning
- having a voice.

FIGURE 2 Providing mobility within the central business district of Ballarat

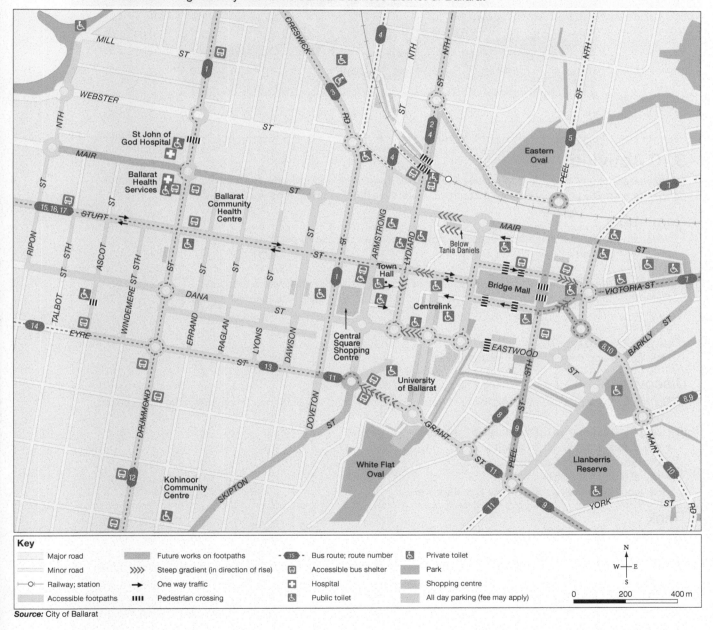

Source: City of Ballarat

Regional cities in Victoria are also conscious of the need to cater for the disabled. The Rural Access Program undertook work in Ballarat to develop a **mobility** map that showed safe, accessible, easy and enjoyable ways to move and connect with the city.

FIGURE 3 Access for everyone, regardless of mobility, is essential.

8.10 Activities

To answer questions online and to receive **immediate feedback** and **sample responses** for every question, go to your learnON title at www.jacplus.com.au. *Note*: Question numbers may vary slightly.

Remember

1. What are the various disabilities that may affect someone's access to public transport?

Explain

2. Explain what you understand by the term accessibility regarding people with a disability.
3. Refer to figure 1. Approximately what percentage of people with disabilities are the same age as:
   (a) you
   (b) your parents
   (c) your grandparents?
   Use figures and percentages in your answer.

Discover

4. Do you know someone with a disability? Explain what types of difficulties they encounter while using public transport.
5. Using the map of the CBD of Ballarat, describe:
   • the accessibility level of Bridge Mall
   • movement around Centrelink
   • travel along Mair Street.
   How would you rank access in the Docklands area?

Think

6. In small groups, devise a trail in a local *environment* for people with disabilities. You might choose your school or a local park, for example. You might also decide what types of disabilities you are planning for, and travel around the site considering the potential needs of the visitor. Where are hazards located? Which areas might the visitor find difficult to navigate? What *places* might be of interest to them? Draw an annotated trail map (only for users who are not sight-impaired) to highlight these various features. To empathise more fully with the needs of others, students could take it in turns to navigate their way around the designated site on crutches, in a wheelchair or blindfolded.

8.11 How do we connect with the world?

Access this subtopic at **www.jacplus.com.au**

8.12 Review

8.12.1 Review

The Review section contains a range of different questions and activities to help you revise and recall what you have learned, especially prior to a topic test.

8.12.2 Reflect

The Reflect section provides you with an opportunity to apply and extend your learning.

Access this subtopic at **www.jacplus.com.au**

TOPIC 9
Tourists on the move

9.1 Overview

Numerous **videos** and **interactivities** are embedded just where you need them, at the point of learning, in your learnON title at www.jacplus.com.au. They will help you to learn the content and concepts covered in this topic.

9.1.1 Introduction

The World Tourism Organization estimates that by 2030, five million people will move each day. Where will these people go and what will influence their choices? What impact will these choices have on the places they visit?

People move to and from many different places for many different reasons.

INQUIRY SEQUENCE

9.1 Overview   158
9.2 What is tourism?   159
9.3 Who goes where?   164
9.4 **SkillBuilder:** Constructing and describing a doughnut chart   online only   166
9.5 Who comes and goes in Australia?   167
9.6 **SkillBuilder:** Creating a survey   online only   170
9.7 **SkillBuilder:** Describing divergence graphs   online only   171
9.8 What are the impacts of tourism?   171
9.9 What can we learn from our travels?   174
9.10 Are zoos and aquariums eco-friendly?   online only   176
9.11 What is cultural tourism?   176
9.12 Is sport a new tourist destination?   179
9.13 **Review**   online only   182

9.2 What is tourism?

9.2.1 Why is tourism important?

The World Tourism Organization defines tourism as the temporary movement of people away from the places where they normally work and live. This movement can be for business, leisure or cultural purposes (see figure 1), and it involves a stay of more than 24 hours but less than one year.

Global tourism increased by 4.7 per cent in 2014: the fifth consecutive year of above average growth. This is despite global economic issues, political change in the Middle East and Africa, and a range of natural disasters. In fact, some of the world's most popular tourist destinations were those affected by global concerns. Figure 2 illustrates why particular destinations are popular. Although tourism

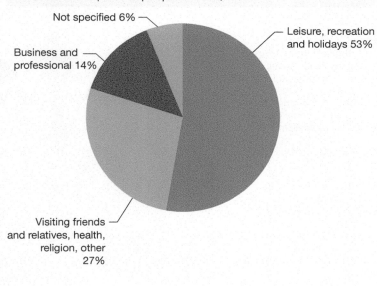

FIGURE 1 Purpose of people's travel, 2014

Not specified 6%
Business and professional 14%
Leisure, recreation and holidays 53%
Visiting friends and relatives, health, religion, other 27%

growth was slower in 2012 (3 per cent), a milestone was reached: there were more than one billion tourist movements in a single year.

Worldwide, 277 million jobs exist because of tourism. Globally, about 9.8 per cent of **gross domestic product (GDP)** is directly linked to the tourism industry, and for many developing countries it is the primary source of income.

FIGURE 2 Types of tourist destinations

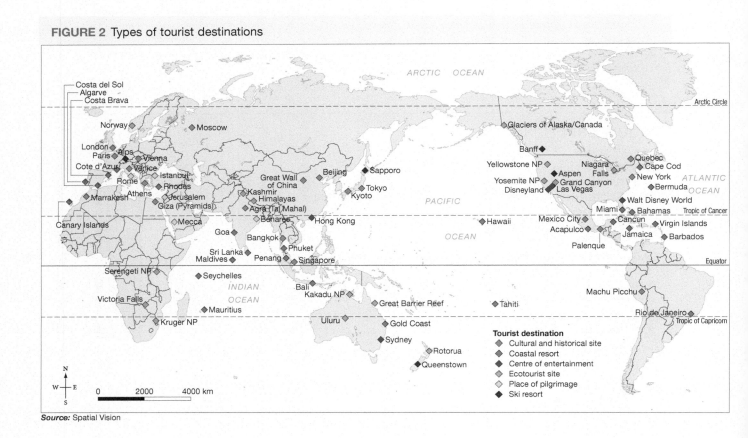

Source: Spatial Vision

FIGURE 3 Four kinds of tourist — are all tourists the same?

Organised mass tourist
- Least adventurous
- Purchases a package with a fixed itinerary
- Does not venture from the hotel complex alone; is divorced from the local community
- Makes few decisions about the holiday

The explorer
- Arranges their own trip
- May go off the beaten track but still wants comfortable accommodation
- Is motivated to associate with local communities and may try to speak the local language

Individual mass tourist
- Similar to the organised mass tourist and generally purchases a package
- Maintains some control over their itinerary
- Uses accommodation as a base and may take side tours or hire a car

The drifter
- Identifies with local community and may live and work within it
- Shuns contact with tourists and tourist hotspots
- Takes risks in seeking out new experiences, cultures and places

9.2.2 Why is tourism increasing?

Tourism has increased dramatically over the past 50 years and continues to grow. Advances in transport technology have reduced not only travel times but also cost.

- Today, you can fly from Australia to Europe in about 20 hours, whereas 40 years ago the same journey took six weeks by boat.
- Today, airline and tour companies offer a range of cut-price deals, and the increased number of competitors for the tourist dollar means that travel is more affordable.
- Increased awareness and knowledge of the world has sparked people's desire to see new places and experience different cultures.
- In general, the travelling public has more leisure time and more disposable income, making both domestic and international travel viable.

9.2.3 What are the latest trends in tourism?

Tourism is an important component in world economies. One in 11 jobs worldwide is linked either directly or indirectly to the tourism industry. In the first 10 years of the twenty-first century, this industry grew by almost 40 per cent, from 674 million international arrivals in 2000 to 1.2 billion in 2014. It is predicted that this figure will rise to 1.8 billion in the year 2030 (see figure 4).

Tourists added US$1.246 billion to the global economy in 2014. Figure 5 shows the top 10 tourism earners for 2014.

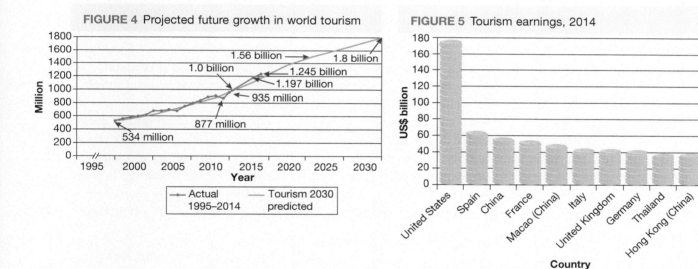

FIGURE 4 Projected future growth in world tourism

FIGURE 5 Tourism earnings, 2014

9.2.4 Factors shaping the future of tourism

Tourism is no longer just for the elite. Improvements in transport and technology have increased our awareness of the world around us. Improved living standards, increased leisure and greater disposable incomes have all created opportunities for people to travel and to experience new places and cultures. These factors are also shaping the tourist of the future (see figure 6). Particular growth areas are the 18–35 market — young people travelling while studying, taking a break from study, or seeing the world before they settle down. The other major area of growth is the over-60 age bracket.

Predictions also suggest that Africa and the Asia–Pacific region will be particular growth areas, attracting more and more of the tourist dollar.

Countries such as Kenya and Tanzania offer a different type of tourist experience. Kenya, for instance, offers:

- safety
- beaches and a tropical climate

- safari parks and encounters with lions and elephants
- a unique cultural experience with the **Masai** people.

The resulting influx of tourists to Kenya has led to the establishment of **national parks** to protect endangered wildlife and promote this aspect of the tourism experience. Money flowing into the region helps improve water quality and infrastructure such as water pipes, roads and airports.

The true challenge for the future, however, is to ensure that:

- money remains in the local economy rather than in the hands of developers, and is used to improve local services, not just tourist services
- the need of indigenous communities to farm the land is balanced with tourist development
- tourist numbers are controlled, to ensure that the environment is not damaged.

FIGURE 6 Factors shaping the future of tourism

9.2.5 Medical tourism

Medical tourism involves people travelling to overseas destinations for medical care and procedures. The low cost of travel, advances in technology and lengthy waiting lists caused by increased demand for elective surgery are turning medical tourism into a multi-billion dollar industry.

While people once travelled overseas only for cosmetic procedures such as facelifts and tummy tucks, the range of services offered has expanded dramatically over recent years to include orthopaedic procedures such as knee and hip replacements, fertility treatments and surrogacy services, and complex heart surgery.

Countries all over the world are attracting patients for a variety of reasons. In some instances, it is the high standard of medical care or the outstanding reputation of a particular facility that attracts people, while for others it is the savings to be made and the opportunity to include a holiday and luxury accommodation as part of the package.

Asia is the market leader in the medical tourism industry, with Thailand and India vying for the number one spot. Thailand is slightly more expensive, but does offer a better tourist experience and has a wider number of services available. India, on the other hand, is cheap and boasts state-of-the-art facilities staffed by Western medical staff, predominantly from the United States. Figure 7 illustrates the savings to be made by having selected medical procedures carried out in Asia rather than Australia; figure 8 shows the savings when a variety of procedures are undertaken in Malaysia as compared to the United States, Thailand or Singapore. With medical tourism expected to add more than $8.5 million to Asian economies per year, it is not surprising that there has been a dramatic increase in the number of facilities to deliver these services.

FIGURE 7 Cost savings that can be made by having medical treatment in Asia versus Australia

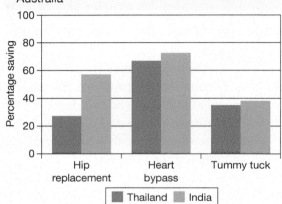

Adapted from: Cosmetic Surgeon India and Rowena Ryan/News.com.au

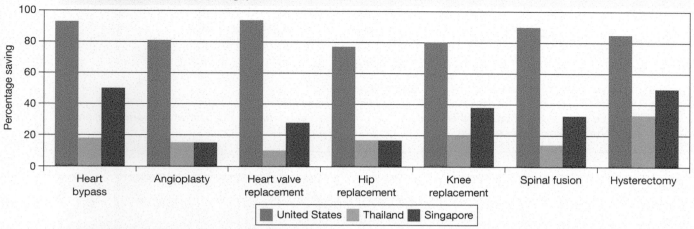

FIGURE 8 Cost savings that can be made by having medical procedures carried out in Malaysia versus the United States, Thailand or Singapore

Adapted from: Cosmetic Surgeon India and Rowena Ryan/News.com.au

9.2 Activities

To answer questions online and to receive **immediate feedback** and **sample responses** for every question, go to your learnON title at www.jacplus.com.au. *Note:* Question numbers may vary slightly.

Remember

1. What is a tourist?
2. Why do you think tourism is one of the fastest growing industries?

Explain

3. Using your atlas as a primary source of information, select three *places* from different categories shown in figure 2 that you might like to visit.
   (a) Calculate the distance between them.
   (b) Explain how you would travel to each *place*.
   (c) Explain what you might expect to see and do in each *place*.
   (d) Work out how long it might take to visit each *place*.
   (e) Describe each location using geographical concepts such as latitude and longitude, direction and *scale*.
   (f) Explain why you have chosen each *place*.

Discover

4. What type of tourist are you? Make a sketch of yourself, similar to the one shown here. Annotate your cartoon to describe yourself as a tourist, using information in this section to help you. Include information about your ideal holiday and explain why you appear as you do in your cartoon.

Predict

5. In 2014, 1.1 billion tourist movements were recorded globally. How many movements is this per day?
6. (a) Using information in this section, predict how many tourists there will be in 2020. What percentage increase does this represent?
   (b) Which *places* do you think will be the most popular?
   (c) What impact do you think these increases will have on the *environment*?
   (d) Will this result in small-*scale* or large-*scale change*?
   (e) Do you think these numbers are *sustainable*? Explain.

7. Tourism expenditure increased by 93 per cent between the year 2000 and the year 2010, from $475 billion to $918 billion. Using these figures as a guide, predict how much income might be generated through tourism by 2030.
8. Look back over your responses to the last three questions. What does this tell you about the importance of tourism? Explain.

Think

9. (a) Explain what you understand by the term *sustainable* tourism.
   (b) Describe an example of tourism that would be considered *sustainable*.
   (c) Describe an example of tourism that would not be considered *sustainable*. Suggest what *changes* might be needed to make it *sustainable*.

Deepen your understanding of this topic with related case studies and questions.
❯ **World tourism**

9.3 Who goes where?

9.3.1 Why is global tourism on the rise?

Tourism is one of the world's largest industries. Even when global economies are experiencing a downturn, people still travel. After natural disasters, countries rely on the return of the tourist dollar to help stimulate their economies. However, the spread of tourism is not uniform across the Earth.

Over time, travel has become faster, easier, cheaper and safer. Economic growth has meant that many people now have more money to spend and can afford to travel. Increases in annual leave have provided people with more time to travel. For example, Australians and New Zealanders can take long-service leave, which is often spent on an extended overseas trip. It has also become more fashionable for young people to spend time seeing the world during their 'gap year' and to travel before settling down and establishing a career and raising a family (see figure 1).

Many young travellers see backpacking as the optimum way to travel. Generally this group:
• is on a tight budget
• wants to mix with other young travellers and local communities
• has a flexible itinerary
• seeks adventure
• is prepared to work while on holiday to extend their stay.

At the other end of the scale, there has also been a dramatic growth in **mature-aged** tourist movements. The number of older people in **developed** countries is growing. In many instances, they have older children who are no longer dependent on their parents for support. Some of these travellers have savings, access to superannuation funds, and the opportunity to retire early; thus, they have both the time and the money to travel.

FIGURE 1 Backpackers spend more, travel further and stay longer than other tourists.

9.3.2 Which countries are top-10 destinations?

As each tourist enters or leaves a country, they are counted by that country's customs and immigration officials. This data is collected by the World Tourism Organization, and the results can be shown spatially (see figure 2).

FIGURE 2 World's top 10 tourist destinations, 2014

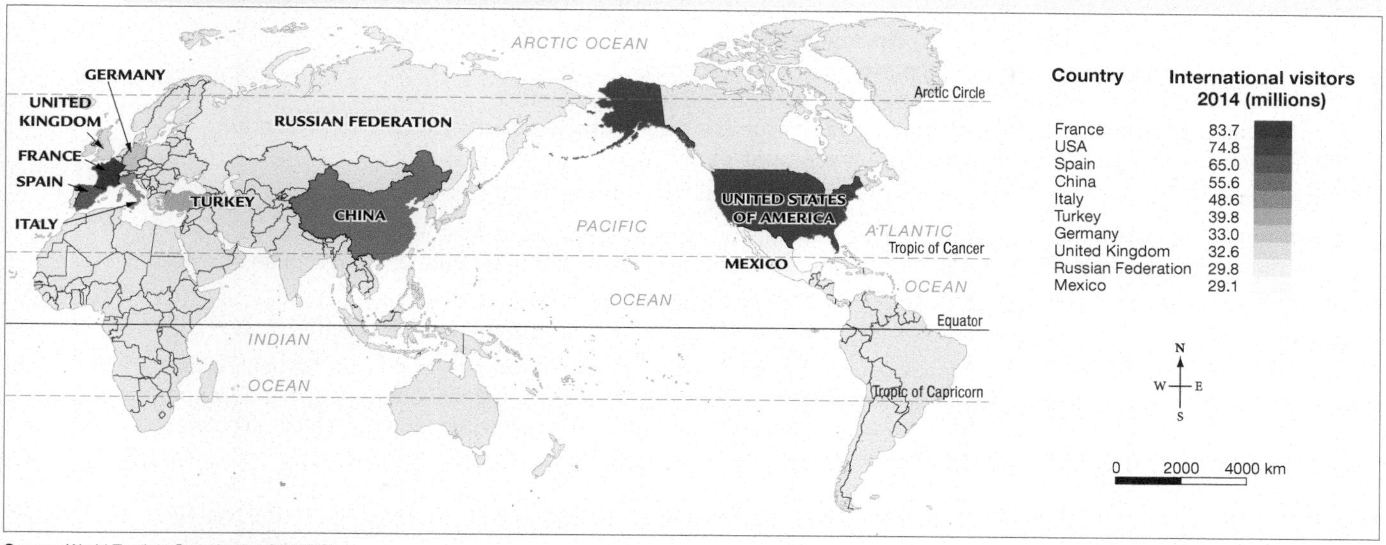

ARCTIC OCEAN
GERMANY
UNITED KINGDOM
FRANCE
SPAIN
ITALY
RUSSIAN FEDERATION
TURKEY
CHINA
PACIFIC
OCEAN
UNITED STATES OF AMERICA
MEXICO
ATLANTIC
OCEAN
INDIAN
OCEAN
Arctic Circle
Tropic of Cancer
Equator
Tropic of Capricorn
Country
International visitors 2014 (millions)
France 83.7
USA 74.8
Spain 65.0
China 55.6
Italy 48.6
Turkey 39.8
Germany 33.0
United Kingdom 32.6
Russian Federation 29.8
Mexico 29.1
N
W
E
S
0 2000 4000 km

Source: World Tourism Organization (UNWTO)

9.3.3 Who spends the most?

Figure 2 shows the countries that attract the most tourists, but which countries do these tourists come from, and how much do they spend? Figure 3 shows the top-10 countries in terms of the money they spend on international tourism.

FIGURE 3 World's top 10 tourist spenders, 2013

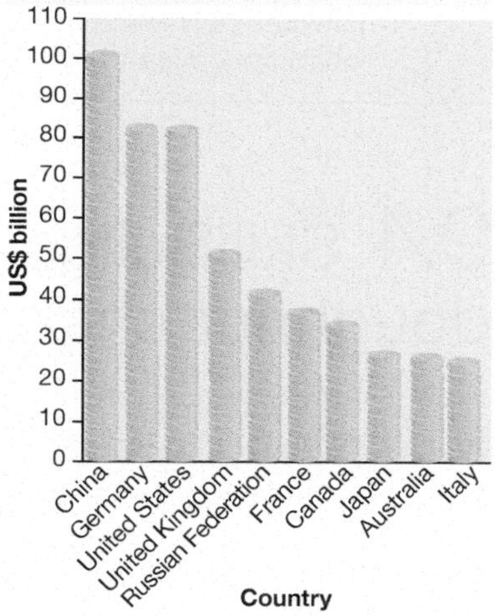

US$ billion
0
10
20
30
40
50
60
70
80
90
100
110
China
Germany
United States
United Kingdom
Russian Federation
France
Canada
Japan
Australia
Italy
Country

9.3.4 Where do people stay?

When travelling overseas, most tourists give little thought to who owns the hotel or resort in which they are staying. Table 1 shows the locations of various hotel-chain headquarters, indicating that the corporate owners of many hotels are based in a country that is often not the one a tourist is visiting. Home Inns entered the market in 2008 when they took over another hotel chain. They are primarily located in China and offer budget accommodation.

TABLE 1 World's top 10 hotel owners, 2015 (based on rooms)

Company	Headquarters (country)	Total hotels	Total rooms
Intercontinental Hotels Group	UK	4840	710 295
Hilton Worldwide	USA	4278	708 268
Marriott International	USA	4117	701 899
Wyndham Hotel Group	USA	7645	660 826
Choice Hotels International	USA	6376	504 808
Accor	France	3717	482 296
Starwood Hotels & Resorts	USA	1207	346 599
Best Western	USA	3900	302 144
Home Inns	China	2609	296 075
Jin Jiang (inc. Louvre Hotels)	China	2208	241 910

9.3 Activities

To answer questions online and to receive **immediate feedback** and **sample responses** for every question, go to your learnON title at www.jacplus.com.au. *Note:* Question numbers may vary slightly.

Explain

1. Carefully study figures 2 and 3 and answer the following.
   (a) On which continents are the top 10 destinations located?
   (b) Which continents are generating the most in tourism spending?
   (c) Describe the *interconnection* between destinations and tourism spending.
2. What is the difference between a mature-age tourist and a backpacker? With the aid of a Venn diagram, show the differences in the needs of these two groups of tourists.

Discover

3. The three main types of tourist attraction are natural, cultural and event attractions. Use a dictionary to help you write your own definition of each term. For each of the countries shown in figure 2, try to find an example of each type of attraction. Use the map in figure 2 in subtopic 9.2 to help you.

Predict

4. (a) On a blank outline map of the world, locate and label the capital cities of each of the top 10 tourist destinations.
   (b) Plot a trip from your nearest capital city to all 10 of these *places*, covering the shortest possible distance, and returning to your capital city. Use the *scale* on the map to estimate the distance travelled. Calculate the time it might take to complete this journey.
5. (a) Describe the pattern of hotel ownership shown in table 1.
   (b) Construct a bar graph showing both the total number of hotels and total number of rooms for the top 10 hotel companies.
   (c) Find out the average cost of a one-week stay in one of these hotel chains.
   (d) How much income would this room generate if it was occupied for each night of the year?
   (e) What percentage of your answer to question 5d would be needed to maintain this room, the complex and other associated costs, such as wages?
   (f) Estimate how much money is left and where you think it goes. What effect do you think this might have on local economies?

9.4 SkillBuilder: Constructing and describing a doughnut chart

WHAT IS A DOUGHNUT CHART?

A doughnut chart is a circular chart with a hole in the middle. Each part of the doughnut is divided as if it were a pie chart with a cut-out. The circle represents the total, or 100 per cent, of whatever is being looked at. The size of the segments is easily seen. Doughnut charts are a useful visual interpretation of data.

Go online to access:

- a clear step-by-step explanation to help you master the skill
- a model of what you are aiming for
- a checklist of key aspects of the skill
- a series of questions to help you apply the skill and to check your understanding.

FIGURE 1 Australia's food export markets by value, 2010–2011

Australia's food export markets by value, 2010-2011

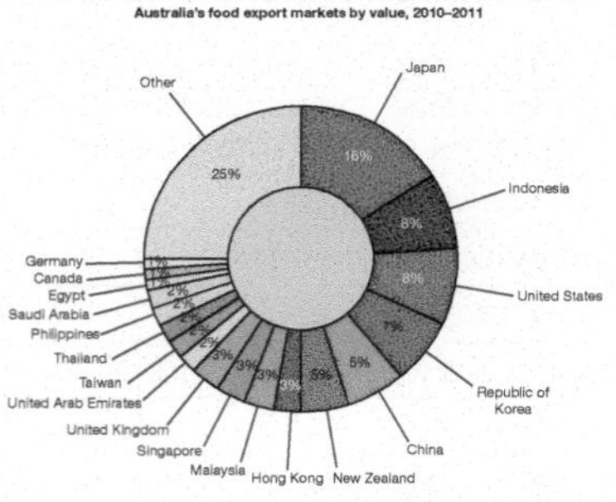

Source: © DAFF 2012, National Food Plan green paper 2012, Department of Agriculture, Fisheries and Forestry, Canberra. CC BY 3.0.

9.5 Who comes and goes in Australia?

9.5.1 Where in the world are Australians going?

Over 13 million Australians go away on holiday each year. Of these, 10 per cent plan to travel overseas, and 55 per cent plan to holiday within Australia. While the numbers of domestic travellers may be high, international tourism is on the rise.

In 2014 Australians made 7.9 international trips and spent a total of 161 million nights abroad. It is estimated that they contributed $47.3 billion to global economies. While the decline in the Australian dollar has seen an increase in the number of Australians choosing to travel domestically we are taking three times as many international holidays today than we were a decade ago.

Despite the decline in the Australian dollar against other currencies in 2014 our currency still has excellent buying power in a wide range of holiday destinations. Competition between airlines; choice of flights; and package deals that include combinations of tours, accommodation, flights and meals continue to fuel the growth of the international market.

The opportunity to live and work overseas has also seen an increase in the number of people under 30 travelling abroad. The under 30s working visa has ensured that foreign travel is both appealing and affordable for this age group. At any one time there are about one million Australians living and working overseas.

Nepal, Japan and the United Arab Emirates are the fastest expanding destinations for Australians travelling abroad.

9.5.2 Who comes here?

In the first eight months of 2015, 6.8 million tourists came to Australia and spent 242 million nights in the country. Approximately 64 per cent of these international visitors had previously been to Australia. They added about $34 billion to the Australian economy. The most visited states were New South Wales with 52 per cent, Queensland with 37 per cent and Victoria with 31 per cent. The countries of origin of these visitors are shown in figure 1, while the reasons for their visits are shown in figure 2.

Nine-hundred and twenty-five thousand jobs can be attributed either directly or indirectly to the tourism industry. It is predicted that by 2020 an additional 123 000 jobs will be created.

9.5.3 What do visitors want to see?

Australia is a land of contrasts, having a wide variety of both human and natural environments. The most popular tourist destinations are shown in figure 3.

FIGURE 1 Country of origin for tourists visiting Australia, and destinations for Australian tourists, 2015

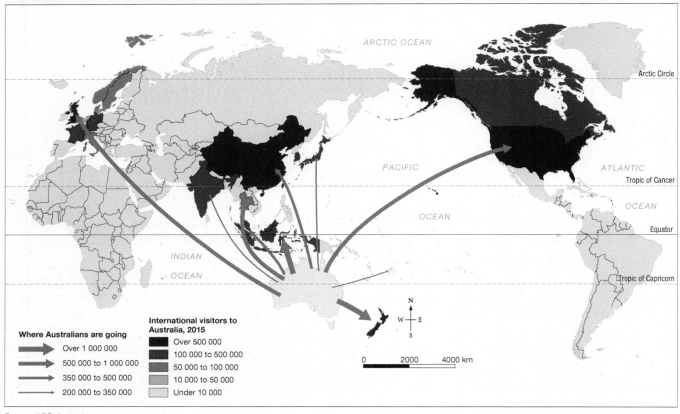

Where Australians are going
→ Over 1 000 000
→ 500 000 to 1 000 000
→ 350 000 to 500 000
→ 200 000 to 350 000

International visitors to Australia, 2015
■ Over 500 000
■ 100 000 to 500 000
■ 50 000 to 100 000
■ 10 000 to 50 000
□ Under 10 000

0   2000   4000 km

Source: ABS, Austrade

FIGURE 2 Reasons for visiting Australia

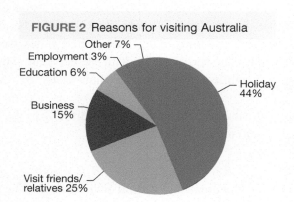

Other 7%
Employment 3%
Education 6%
Business 15%
Holiday 44%
Visit friends/relatives 25%

TABLE 1 How we see ourselves

New South Wales	Nightlife
Northern Territory	Outback
Queensland	Beaches
South Australia	Wine
Tasmania	Nature
Victoria	Sport
Western Australia	Wine

FIGURE 3 Australia's most popular tourist destinations

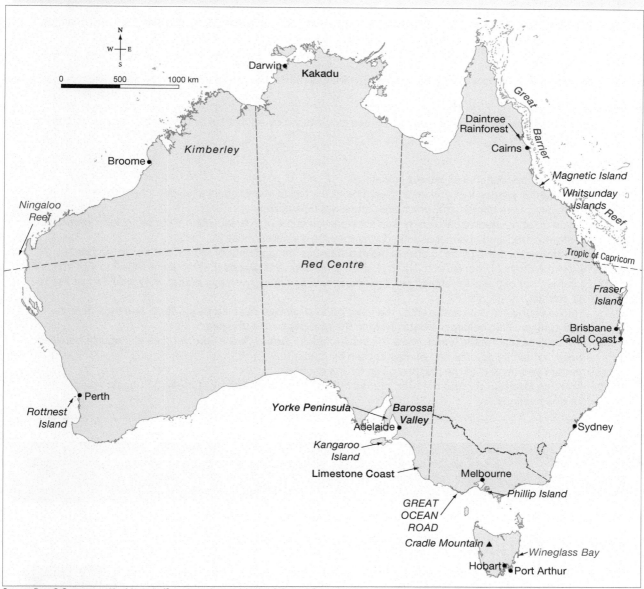

Source: Data © Commonwealth of Australia (Geoscience Australia) 2013 & © State of Queensland (Department of Agriculture, Fisheries and Forestry) 2013

How are we perceived?

Tourism statistics have revealed that Victoria has increased its share of the international tourist market. Tourism in this state grew by 28 per cent in the 12-month period ending September 2015, a figure that is significantly higher than the national average, which is just 15 per cent. New South Wales, however, remains the most visited state.

9.5 Activities

To answer questions online and to receive **immediate feedback** and **sample responses** for every question, go to your learnON title at www.jacplus.com.au. *Note*: Question numbers may vary slightly.

Remember

1. Why might more Australians choose to holiday overseas rather than in Australia?

Explain

2. Explain the *interconnection* between the *places* most visited by Australians and our major source of tourists.

Discover

3. Study figure 3.
   (a) Identify the top tourist destinations in Victoria.
   (b) Suggest other *places* that you think should be at the top of every tourist's holiday itinerary.
   (c) Identify which are human and which are natural *environments*.
   (d) Prepare an annotated visual display that showcases places of interest within Victoria. Include information about the attractions, their location and why they are a viewing must.
   (e) What strategies are in place to ensure the *sustainable* management of tourist facilities in Victoria?
4. (a) Survey members of your class to find out which three overseas *places* they would most like to visit and why.
   (b) Each member of the class should ask their parents which three overseas *places* they would most like to visit and why.
   (c) Compile your class data and identify the most popular *places* selected by students and their parents. Make sure you also collate the data showing the reasons for the choices.
   (d) On an outline map of the world, show the results of your survey. Make sure you can distinguish between *places* chosen by parents and *places* chosen by students.
   (e) Annotate your map with the reasons given for the choices.
   (f) Is there an *interconnection* between *places* chosen by parents and by students? Suggest reasons for your observations.

myWorldAtlas Deepen your understanding of this topic with related case studies and questions.
⊙ **Tourism in Australia**

9.6 SkillBuilder: Creating a survey

WHAT IS A SURVEY?

Surveys collect primary data. A survey involves asking questions, recording and collecting responses, and collating and interpreting the number of responses. Surveys are useful because they provide statistics for a specific topic that might not be available by any other means. A wide range of data can be gathered in an efficient and simple way.

Go online to access:
- a clear step-by-step explanation to help you master the skill
- a model of what you are aiming for
- a checklist of key aspects of the skill
- a series of questions to help you apply the skill and to check your understanding.

FIGURE 1 A survey of shoppers

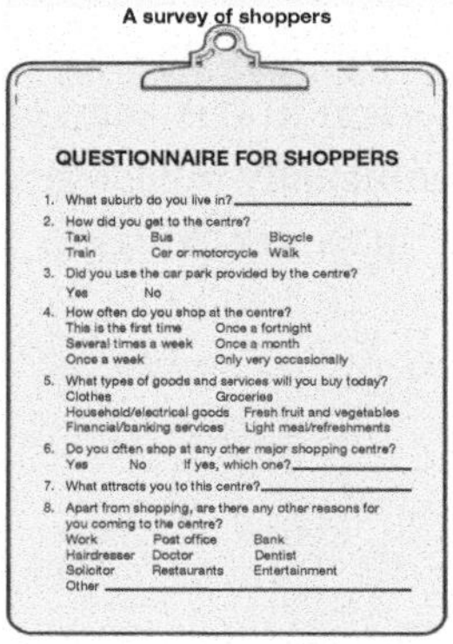

A survey of shoppers

QUESTIONNAIRE FOR SHOPPERS

1. What suburb do you live in?
2. How did you get to the centre?
   Taxi   Bus   Bicycle
   Train   Car or motorcycle   Walk
3. Did you use the car park provided by the centre?
   Yes   No
4. How often do you shop at the centre?
   This is the first time   Once a fortnight
   Several times a week   Once a month
   Once a week   Only very occasionally
5. What types of goods and services will you buy today?
   Clothes   Groceries
   Household/electrical goods   Fresh fruit and vegetables
   Financial/banking services   Light meals/refreshments
6. Do you often shop at any other major shopping centre?
   Yes   No   If yes, which one?
7. What attracts you to this centre?
8. Apart from shopping, are there any other reasons for you coming to the centre?
   Work   Post office   Bank
   Hairdresser   Doctor   Dentist
   Solicitor   Restaurants   Entertainment
   Other

9.7 SkillBuilder: Describing divergence graphs

WHAT IS A DIVERGENCE GRAPH?

online only

A divergence graph is a graph that is drawn above and below a zero line. Those numbers above the line are positive, showing the amount above zero. Negative numbers that are shown indicate that the data has fallen below zero. A divergence graph allows you to identify changes away from the norm in a trend.

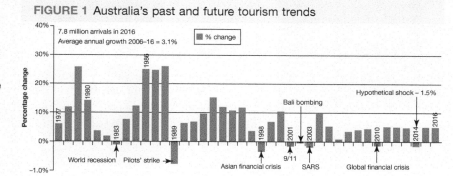

FIGURE 1 Australia's past and future tourism trends

Go online to access:

- a clear step-by-step explanation to help you master the skill
- a model of what you are aiming for
- a checklist of key aspects of the skill
- a series of questions to help you apply the skill and to check your understanding.

myWorldAtlas **Deepen your understanding of this topic with related case studies and questions.**
 ❂ **Tourism in Australia**

9.8 What are the impacts of tourism?

9.8.1 Do the benefits outweigh the costs?

Tourism seems like the perfect industry. It can encourage greater understanding between people and bring prosperity to communities (see figure 2). However, tourism development can also destroy people's culture and the places in which they live (see figure 1). There is sometimes a fine line between exploitation and sustainable tourism.

FIGURE 1 The negative impacts of tourism

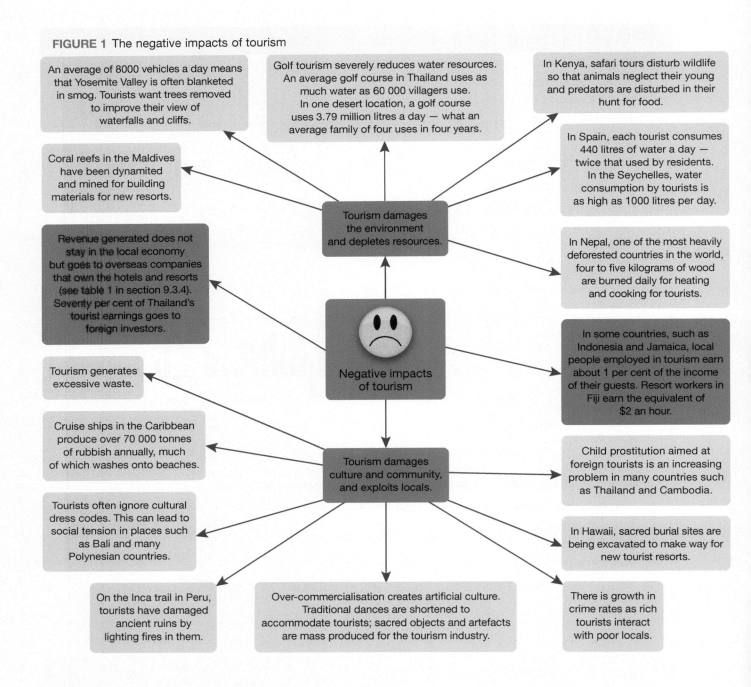

An average of 8000 vehicles a day means that Yosemite Valley is often blanketed in smog. Tourists want trees removed to improve their view of waterfalls and cliffs.

Golf tourism severely reduces water resources. An average golf course in Thailand uses as much water as 60 000 villagers use. In one desert location, a golf course uses 3.79 million litres a day — what an average family of four uses in four years.

In Kenya, safari tours disturb wildlife so that animals neglect their young and predators are disturbed in their hunt for food.

Coral reefs in the Maldives have been dynamited and mined for building materials for new resorts.

In Spain, each tourist consumes 440 litres of water a day — twice that used by residents. In the Seychelles, water consumption by tourists is as high as 1000 litres per day.

Revenue generated does not stay in the local economy but goes to overseas companies that own the hotels and resorts (see table 1 in section 9.3.4). Seventy per cent of Thailand's tourist earnings goes to foreign investors.

In Nepal, one of the most heavily deforested countries in the world, four to five kilograms of wood are burned daily for heating and cooking for tourists.

Tourism damages the environment and depletes resources.

Negative impacts of tourism

Tourism generates excessive waste.

In some countries, such as Indonesia and Jamaica, local people employed in tourism earn about 1 per cent of the income of their guests. Resort workers in Fiji earn the equivalent of $2 an hour.

Cruise ships in the Caribbean produce over 70 000 tonnes of rubbish annually, much of which washes onto beaches.

Tourism damages culture and community, and exploits locals.

Child prostitution aimed at foreign tourists is an increasing problem in many countries such as Thailand and Cambodia.

Tourists often ignore cultural dress codes. This can lead to social tension in places such as Bali and many Polynesian countries.

In Hawaii, sacred burial sites are being excavated to make way for new tourist resorts.

On the Inca trail in Peru, tourists have damaged ancient ruins by lighting fires in them.

Over-commercialisation creates artificial culture. Traditional dances are shortened to accommodate tourists; sacred objects and artefacts are mass produced for the tourism industry.

There is growth in crime rates as rich tourists interact with poor locals.

FIGURE 2 The positive impacts of tourism

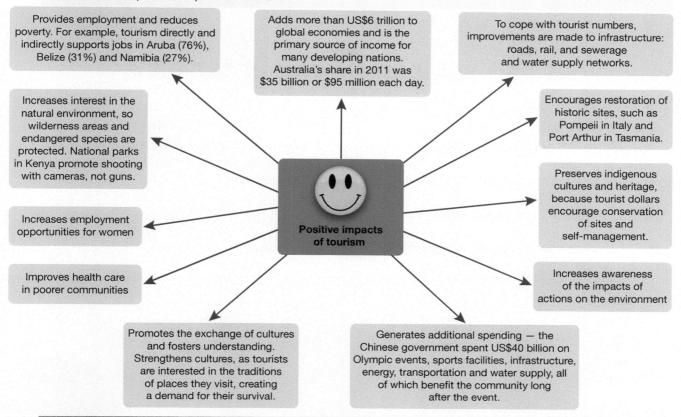

Provides employment and reduces poverty. For example, tourism directly and indirectly supports jobs in Aruba (76%), Belize (31%) and Namibia (27%).

Adds more than US$6 trillion to global economies and is the primary source of income for many developing nations. Australia's share in 2011 was $35 billion or $95 million each day.

To cope with tourist numbers, improvements are made to infrastructure: roads, rail, and sewerage and water supply networks.

Increases interest in the natural environment, so wilderness areas and endangered species are protected. National parks in Kenya promote shooting with cameras, not guns.

Encourages restoration of historic sites, such as Pompeii in Italy and Port Arthur in Tasmania.

Increases employment opportunities for women

Preserves indigenous cultures and heritage, because tourist dollars encourage conservation of sites and self-management.

Improves health care in poorer communities

Positive impacts of tourism

Increases awareness of the impacts of actions on the environment

Promotes the exchange of cultures and fosters understanding. Strengthens cultures, as tourists are interested in the traditions of places they visit, creating a demand for their survival.

Generates additional spending — the Chinese government spent US$40 billion on Olympic events, sports facilities, infrastructure, energy, transportation and water supply, all of which benefit the community long after the event.

9.8 Activities

To answer questions online and to receive **immediate feedback** and **sample responses** for every question, go to your learnON title at www.jacplus.com.au. *Note*: Question numbers may vary slightly.

Think

1. Using figures 1 and 2, explain how tourism can have both positive and negative effects.
2. (a) Figure 3 shows how the tourist dollar can flow from one job to the next. Those jobs in the centre of the diagram interact directly with the tourist, while those on the outside do not. Copy the diagram into your workbook at an enlarged size. Complete it by adding other jobs. Add to it if you can. Study your completed diagram and write a paragraph explaining the *interconnection* between tourism and the economy.
   (b) Repeat this exercise looking at either the social or *environmental* impacts.

FIGURE 3 One view of tourism

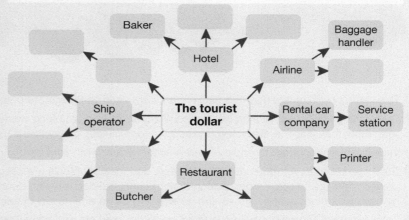

Baker

Hotel

Baggage handler

Airline

Ship operator

The tourist dollar

Rental car company

Service station

Printer

Restaurant

Butcher

3. The type of *interconnection* shown between industries in figure 3 is sometimes called the multiplier effect. Explain what you think this means.
4. The impact of tourism can be classified as *environmental*, cultural and economic. Study figure 1, showing the negative impacts of tourism. Working in groups of three, select an impact from each group. Explain the *scale* of each impact and devise a strategy for *sustainable* tourism.
5. Which of the following would be the best to develop as a tourist resource in your region: art gallery, museum, cinema complex or sports stadium? Justify your answer.

9.9 What can we learn from our travels?

9.9.1 What is ecotourism?

Tourism has the capacity to benefit environments and cultures or destroy them. **Ecotourism** has developed in response to this issue. The aim is to manage tourism in a sustainable way. This might be through educational programs related to the environment or cultural heritage, or through controlling the types and locations of tourist activities or the number of tourists visiting an area.

FIGURE 1 Anatomy of an ideal ecotourism resort

Ⓐ The natural bush is retained and native plants are used to revegetate or landscape the area.

Ⓑ Composting toilets treat human waste, and worm farms consume food waste. Water is treated with ultraviolet light rather than chlorine. Recycling is practised; for example, grey water is used in irrigation and toilet systems.

Ⓒ Visitors are encouraged to improve and maintain the environment by using paths or planting trees.

Ⓓ Buildings blend in with the natural landscape, and local materials are used. Buildings are often raised to prevent damage to plant roots. During construction, builders prevent contamination of the local environment by having workers change shoes and by washing down equipment to keep out foreign organisms.

Ⓔ Local organically grown produce is used, and craft markets and stalls might also be established and run by indigenous communities, supporting the local economy, creating jobs and reducing poverty.

Ⓕ There is no golf course, because of the high water use and need for pesticides it would require.

Ⓖ Low-impact, non-polluting transport such as bicycles are provided for guests.

Ⓗ Walking trails include educational information boards.

I An information centre helps visitors understand the environment. Local indigenous people are employed to educate visitors about their culture.

J Electricity is generated through solar panels on the roofs of eco-cabins.

K Boardwalks are built over sensitive areas such as sand dunes to protect them from damage. Boardwalks might also be constructed in the tree canopies.

L Trained guides educate tourists about coral reefs and native vegetation, and show visitors how to minimise their impact.

Ecotourism differs from traditional tourism in two main ways.

• It recognises that many tourists wish to learn about the natural environment (such as reefs, rainforests and deserts) and the cultural environment (such as indigenous communities).

• It aims to limit the impact of tourist facilities and visitors on the environment.

Ecotourism is the fastest growing sector in the tourism industry, increasing at about 10 to 15 per cent per year.

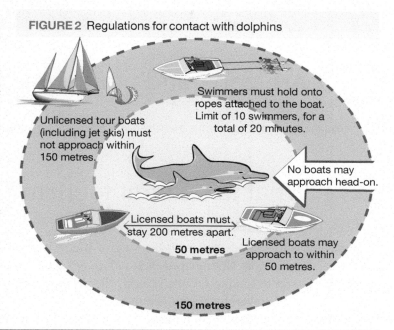

FIGURE 2 Regulations for contact with dolphins

Unlicensed tour boats (including jet skis) must not approach within 150 metres.

Swimmers must hold onto ropes attached to the boat. Limit of 10 swimmers, for a total of 20 minutes.

No boats may approach head-on.

Licensed boats must stay 200 metres apart.

50 metres

Licensed boats may approach to within 50 metres.

150 metres

9.9 Activities

To answer questions online and to receive **immediate feedback** and **sample responses** for every question, go to your learnON title at www.jacplus.com.au. *Note*: Question numbers may vary slightly.

Remember

1. How does an ecotourism resort differ from a traditional tourist resort?

Explain

2. Use a mind map to explore further *changes* that could be made to the resort shown in figure 1 to make it even more *environmentally* friendly. Would you describe your *changes* as small-*scale* or large-*scale*? Justify how these *changes* might be more *environmentally sustainable*.

Discover

3. Visitors to ecotourism resorts are often attracted by brochures that emphasise the resort's *environmental* policies. These brochures also set out guidelines to follow in order to minimise visitor impact.
   (a) Design and produce a brochure for the ecotourism resort illustrated in figure 1. Use ICT tools and techniques to maximise the brochure's impact.
   (b) Add another eco-activity to the island and devise strategies to educate tourists and minimise their impact on the *environment*.

Think

4. One of the most famous examples of wildlife-based ecotourism in Australia is Monkey Mia in Western Australia. Here the wild dolphins come into shore and tourists are able to feed, swim with and touch them.
   (a) What rules and other techniques are used to control the interaction between dolphins and tourists?
   (b) Predict potential problems that might occur between dolphins and tourists.
   (c) Do you think this is an example of *sustainable* ecotourism? Give reasons for your answer.

 myWorldAtlas **Deepen your understanding of this topic with related case studies and questions.**
- ❂ **Kakadu National Park**
- ❂ **Wilsons Promontory**
- ❂ **Ningaloo Reef and ecotourism**

9.10 Are zoos and aquariums eco-friendly?

 on**line** only

To access this subtopic, go to your learnON title at **www.jacplus.com.au**

9.11 What is cultural tourism?

9.11.1 Different cultures

Cultural tourism is concerned with the way of life of people in a geographical region. It is usually connected to elements that have shaped their values or culture, such as a shared history, traditions or religion. It may be linked to unique events, such as the celebration of Chinese New Year or Thanksgiving in the United States. Whatever the reason, the mass movement of people associated with these events has a significant impact on both people and places.

9.11.2 What is the impact of Thanksgiving?

Thanksgiving is held in the United States on the fourth Thursday in November. It dates back to the seventeenth-century celebration of the harvest. Today it is a time for families to get together and give thanks for what they have.

The holiday period runs from Wednesday to Sunday. In 2015, about 46.9 million people travelled an average distance of 884 kilometres to celebrate with family and friends (see figure 1). As millions of people travel across the United States, transport systems are stretched to the limits, creating delays and traffic congestion. Because the holiday season is so close to the start of winter, the weather can further complicate people's travel plans, especially those who live in the northern states. Early winter storms can bring ice and snow, resulting in airport closures and impassable roads.

Surprisingly, in 2015 the cost of travel decreased around the time of Thanksgiving. Travellers paid less for fuel, car hire, accommodation and airfares.

9.11.3 What is the impact of Chinese New Year?

Chinese New Year is the longest and most important of the traditional Chinese holidays. Dating back centuries, it is steeped in ancient myths and traditions. The festivities begin on the first day of the first month in the traditional Chinese calendar, and last for 15 days. They conclude with

FIGURE 1 Thanksgiving and modes of transport

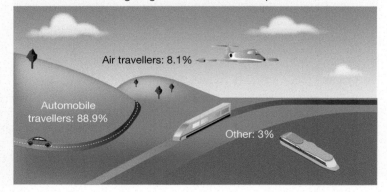

Air travellers: 8.1%

Automobile travellers: 88.9%

Other: 3%

the lantern festival on Chinese New Year's Eve, a day when families gather for their annual reunion dinner. It is considered a major holiday, and it influences not only China's geographical neighbours but also the nations with whom China has economic ties.

Of special significance is the fact that the date on which Chinese New Year occurs varies from year to year. This date coincides with the second **new moon** after the Chinese **winter solstice**, which can occur any time between 21 January and 20 February.

Chinese New Year, or Lunar New Year, is celebrated as a public holiday in many countries with large Chinese populations or with calendars based on the Chinese lunar calendar (see figure 2). The changing nature of this holiday has meant that many governments have to shift working days to accommodate this event.

In China itself, many manufacturing centres close down for the 15-day period, allowing tens of millions of people to travel from the industrial cities where they work to their hometowns and rural communities. This means that retailers and manufacturers in overseas countries such as the United States and Australia have to adjust their production and shipping schedules to ensure they have enough stock on hand to deal with the closure of factories in China. For those shopping online, goods simply will not be available and will be placed on back order.

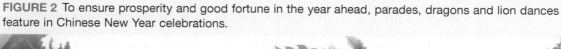

FIGURE 2 To ensure prosperity and good fortune in the year ahead, parades, dragons and lion dances feature in Chinese New Year celebrations.

9.11.4 The logistics of moving 80 million people each day

Chinese New Year has been described as the biggest annual movement of people in China. Over a five-day period, an average of 80 million journeys are recorded in the last-minute dash to make it home for the traditional family celebrations. If you include the 40 days surrounding 8 February (the date of Chinese New Year in 2016), more than 3.65 billion people travel in China alone.

Most people elect to travel by road as most middle-class citizens cannot afford to fly home, nor do they want to queue for hours or days to purchase bus or rail tickets. Nevertheless, in 2016, 24 million chose to travel by air; in fact, one airline added an additional 210 flights to cope with the increased demand. However, weather conditions and the impact of additional flights competing for the same amount of air space meant that about 80 per cent of flights were delayed. Delays of more than five hours were not uncommon. It is also not uncommon for commuters to add thousands of kilometres to their journey. One airline traveller flew from Beijing to Kunming in southern China via Bangkok because he could not get a direct flight.

A growing trend in both 2015 and 2016 was motorcycle convoys as these vehicles are not only cheaper to run but have the added advantage of being able to avoid the congestion created by 80 million people and their vehicles clogging the motorways.

Rail travel was not without its own problems as a system with a capacity to move 3.4 million people had to cope with 145 million travellers. In 2012 a new online booking system was introduced; however, it crashed when 1.66 million people attempted to log on at the same time. The system is still plagued with issues and unable to keep pace with the increased demand and has done little to address the issue of scalpers. Even with extended selling hours people were faced with long queues that often saw tempers flare. Prices were also often double or triple the usual cost as scalpers cashed in on those desperate to get home. In 2015 the rail network was expanded through the addition of 9000 kilometres of track, making it the world's second largest rail network, second only to the United States.

In 2016 almost 100 000 people were left stranded at railway stations after ice and snow in other parts of the country caused long delays. Fifty-five trains in Shanghai and 24 in Guangzhou were unable to leave their respective stations when China was struck by a record-breaking cold snap. Almost 4000 police and security guards were called in to keep order (see figure 3).

FIGURE 3 Travel chaos as crowds swell outside Guangzhou station after bad weather causes long delays

9.11 Activities

To answer questions online and to receive **immediate feedback** and **sample responses** for every question, go to your learnON title at www.jacplus.com.au. *Note*: Question numbers may vary slightly.

Explain

1. In your own words, explain what is meant by the term *cultural tourism*.
2. Why are Thanksgiving and Chinese New Year regarded as cultural events?

Discover

3. As a class, brainstorm a list of cultural or celebratory events that occur in Australia.
4. Use the internet to find out more about either Chinese New Year or Thanksgiving. Investigate the history, myths and traditions associated with your chosen event. Prepare an annotated visual display comparing your finding with a cultural or celebratory event in Australia. Make sure you include references to the *scale* of your chosen event and the *place* in which it occurs.

Think

5. Copy the table below into your workbook and fill it in. Use the **Thanksgiving** weblink in the Resources tab to find out more and help you complete your table.
   (a) What is the preferred mode of transport for Thanksgiving and for Chinese New Year? Suggest reasons for differences in travel arrangements. In your response, include reference to the *scale* of movement.
   (b) Make a list of problems associated with the mass movement of people.
   (c) Select one of the problems you have identified and explain the impact it might have on people, *places* and the *environment*. Suggest a strategy for the *sustainable* management of this problem in order to reduce its impact.
6. Write a paragraph explaining how cultural events can *change* people, *places* and the *environment*.

	Thanksgiving	Chinese New Year
Number of trips		
Most common form of transport		
Length of holiday period		
Purpose of trip/activities		

learn on RESOURCES — ONLINE ONLY

Explore more with this weblink: Thanksgiving

9.12 Is sport a new tourist destination?

9.12.1 How are tourism and sport connected?

Sport tourism involves people travelling to view or participate in a sporting event. It is an expanding sector of the tourism industry, currently adding $600 million to global economies each year. It is estimated that, on average, 12 million international trips are made to view sporting events. But what impact does this have on people and places?

Governments spend millions of dollars to attract people to sporting events such as the Olympics, the cricket and soccer world cups, and motor racing events, to name just a few. These events also trigger:

- construction of new stadiums
- expansion and upgrades of transport networks
- improvements to airport facilities
- clean-ups of cities in readiness for the arrival of tourists.

Sports tourists fall into two broad categories: those who like to watch and those who want to participate. The latter view sport as a part of their leisure and recreational activities (see figure 1). A common trait in all sports tourists is their passion and willingness to spend money to indulge this passion.

9.12.2 Are the Olympics a tourist bonanza?

Major sporting events such as the Olympics translate into improved infrastructure, and provide the host city with considerable international exposure. Does this bring in more tourists and justify the capital outlay? (See figure 2.)

The general consensus is that the costs associated with hosting a one-off major event, such as the soccer World Cup or the Olympic Games, do not meet the anticipated outcomes. Statistics show that three million tourists visited the United Kingdom in August 2012 (their Olympic year) — 5 per cent *less* than in the previous year. Tourism spending, however, went up by 9 per cent, in part because of spending on Olympic tickets. In addition, many UK residents chose to holiday overseas rather than remain at home during the Olympic Games. Organisers were also frustrated by the number of empty seats in many of the venues. On the plus side, however, building the Olympic village provided a £6 billion boost to the building and construction industry.

FIGURE 1 On the trail — recreational tourism

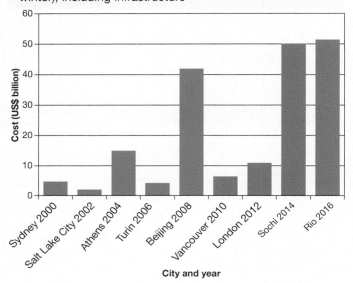

FIGURE 2 Olympics expenditure (both summer and winter), including infrastructure

But what happens to the people who originally lived on the site of the proposed new venues and athletes' village? Quite simply, they are moved on. While they may receive some compensation, land values go up in the shadow of renewed development. Residents simply cannot afford to live in the new developments, nor can they afford to renovate their existing dwelling. In the lead-up to the Beijing Olympics, 1.5 million Chinese people were forced out of their homes to make way for Olympics venues.

Once the event is over, many of the stadiums are under-used, and it can take years to recover from the cost of staging the event. The city of Montreal, for instance, took 30 years to pay back the equivalent of US$6 billion (in today's money) in Olympic spending.

FIGURE 3 The opening ceremony of the London Olympic Games

9.12.3 What is the status of other sports events?

It has generally been accepted that regular sporting events can have financial benefits. Many international tourists visiting the United Kingdom include a sporting event on their itinerary. Most popular is soccer, because of the opportunity to see some of the world's most talented athletes playing in some the UK's top teams. In 2014, 800 000 international fans added £684 million to the British economy. Official figures show that these fans spent an average of £850 — far more than the average £570 spent by the rest of the tourist population.

Overall, sports tourists stay longer and are not deterred by the weather. The popularity of football is also evident in Australia, where three separate codes attract huge crowds every week, and fans are prepared to travel interstate to watch their team play.

FIGURE 4 The Barmy Army comprises thousands of fans who come to Australia to cheer on the English cricket team.

But it is not just football that attracts the crowds. The English cricket team is followed around the world by its unofficial cheer squad, nicknamed the Barmy Army. Other 'fanatics' based in Australia organise 58 tours to 57 destinations each year, taking in some of the biggest sporting events both at home and abroad.

9.12 Activities

To answer questions online and to receive **immediate feedback** and **sample responses** for every question, go to your learnON title at www.jacplus.com.au. *Note:* Question numbers may vary slightly.

Remember

1. (a) Is someone who goes to a local football match a tourist? Explain.
   (b) What if that person travels interstate? Explain.

Explain

2. Brainstorm a list of sports that might spend money to attract tourists. Categorise these as **hard** and **soft sport tourism** events.
3. Compile a table that highlights the positives and negatives of sport tourism. Choose two positives and two negatives. For each, explain the impact it has on people and *places*.

Think

4. The Phillip Island Grand Prix racing circuit is located 100 kilometres south of Melbourne, Victoria. The island is linked to the mainland by a bridge. The area is popular for its beaches and wildlife, but it is also home to a racing circuit that stages a variety of motor sports throughout the year. Collectively, more than $110 million is generated annually from the circuit's car and bike activities. Three events — the Moto GP, V8 Supercars and Superbikes — bring in over $79 million. Each of these events brings more than 65 000 people to the island.
   (a) What facilities would be needed to cater for such a large influx of people?
   (b) Figure 3 in subtopic 9.8 shows how the tourist dollar can flow from one job to the next. Complete a diagram like this for the Phillip Island Grand Prix circuit.
   (c) With a partner, brainstorm a list of negative consequences that might result from having a Grand Prix circuit on Phillip Island. Make sure you consider the impact on people and the *environment*, as well as the *scale* of such effects.
   (d) Write a paragraph explaining the *interconnection* between the location of sporting facilities and their impact on people and *places*.
   (e) Do you think this is an example of *sustainable* tourism? Justify your point of view.

 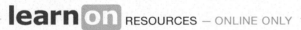 RESOURCES — ONLINE ONLY

 **Try out this interactivity:** Are the Olympic Games worth gold? (int-3337)

---

my**World**Atlas **Deepen your understanding of this topic with related case studies and questions.**
   ❍ **Sport**
   ❍ **The FIFA World Cup**

9.13 Review

9.13.1 Review

The Review section contains a range of different questions and activities to help you revise and recall what you have learned, especially prior to a topic test.

9.13.2 Reflect

The Reflect section provides you with an opportunity to apply and extend your learning.

Access this subtopic at **www.jacplus.com.au**

TOPIC 10
Buy, swap, sell and give

10.1 Overview

Numerous **videos** and **interactivities** are embedded just where you need them, at the point of learning, in your learnON title at www.jacplus.com.au. They will help you to learn the content and concepts covered in this topic.

10.1.1 Introduction

Trade, in the form of buying, swapping, selling and giving goods and services, is a driving force that interconnects people and places all over the world. Trade has gone on ever since human societies existed. In contrast, international aid is a modern phenomenon, although countries have always had internal programs to help those in need. Trade and aid can bring people together to share the Earth's resources, but there can be problems when those resources are limited and, potentially, negative consequences for the environment. The big question is how to organise trade and aid so that it fosters social justice and is fair and sustainable.

Container terminals, Hong Kong, China

Starter questions

1. What goods and services do you need to support your lifestyle? Think about your everyday life at home, at school, and in sport, recreation and hobbies.
2. Where do you access your goods and services? To what extent do you obtain goods and services online?
3. From what you know, do you think all trade that occurs in the world today is fair? How can fair trade and aid make the world a better *place* for all?

INQUIRY SEQUENCE

10.1 Overview		183
10.2 How does trade connect us?		184
10.3 How does trade connect Australia with the world?		187
10.4 How is food traded around the world?		191
10.5 **SkillBuilder:** Constructing multiple line and cumulative line graphs	online only	194
10.6 How has the international automotive trade changed?		195
10.7 Are global players altering the industrial landscape?		197
10.8 Why is fair trade important?		200
10.9 Why does Australia give foreign aid?		203
10.10 Why is the illegal wildlife trade a cause for concern?	online only	206
10.11 **SkillBuilder:** Constructing and describing a flow map	online only	206
10.12 Review	online only	207

10.2 How does trade connect us?

10.2.1 Trade in goods and services

The Earth's resources are not distributed evenly over space. For instance, some places may have an abundance of iron ore and others may have none. To solve this problem, nations have developed trade, allowing producers and consumers to exchange goods and services.

The system of trade has been around for a long time. Its earliest form was bartering at local markets or fairs. Merchants also used land and sea routes to access markets in foreign lands, where they exchanged goods for payment. More recently, air transport has become a means of trade, and the internet has made it possible to instantly exchange information. Today, we have a highly sophisticated, large-scale, global system of trade.

A modern example of the interconnection of trade is the production of the Airbus A380. To construct this plane, component parts must be purchased from different countries and transported over land and sea to reach their final assembly place in Toulouse, France (see figure 1).

Goods and services, of which there are many, are generated by either processing Earth's resources or people doing things for each other.

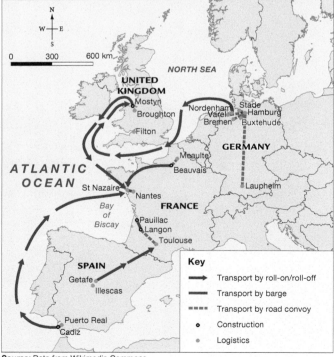

FIGURE 1 The component parts routes of the Airbus A380

Source: Data from Wikimedia Commons.

A good can be an item as simple as a loaf of bread or it can be as complex as a motor car.

A service is not something you can hold in your hand; examples of services are education in a school or the advice a doctor gives a patient. What types of goods and services do you use to support your lifestyle?

As seen in figure 2, the processing of a resource into more complex goods can be a series of transitions, in which there is **value adding** at each level of industry (that is, its value increases). An important consideration in the production of goods and services is the impact on the environment.

10.2.2 How are goods and services consumed?

Household final consumption per person

Household final consumption per person is the value or money spent on all goods and services such as food, cars, washing machines, electricity, water and gas, education, medical service expenses and entertainment. The total figure of all goods and services when calculated for a country for a year is then divided by the total population of the country to give a figure in dollars and this is then referred to as the household final consumption per person.

The highest consumers of goods and services on a per-person basis are wealthy, industrialised countries such as the United Arab Emirates, the United States and Hong Kong (see figure 3). However, countries such as China and India consume high levels of goods and services because they have very large populations. As would be expected, countries that are high-level consumers can have significant impacts on the environment, particularly in terms of energy use and waste production.

What about the poorest countries in the world?

At the lower end of the scale of household final consumption per person, people in countries such as Malawi and Burundi spend $507 and $468 respectively per year, or less than $1.40 per day per person. This expenditure is mainly for food (see figure 4).

FIGURE 2 Four levels of industry

Primary industry
Takes natural resources from the Earth or grows them
Example: a farmer grows corn that is then transported to a canning factory.

Secondary industry
Makes products from natural resources
Example: a factory makes tins of corn and sells them to supermarkets.

Tertiary industry
Sells products or services
Example: a supermarket sells tins of corn and other products to consumers.

Quaternary industry
Sells knowledge and information
Example: a marketing analyst works out how best to position products, and sells this information to supermarkets.

FIGURE 3 Top 13 countries for household final consumption, per person. Australia is ranked twelfth in this list for the years 2013–2014

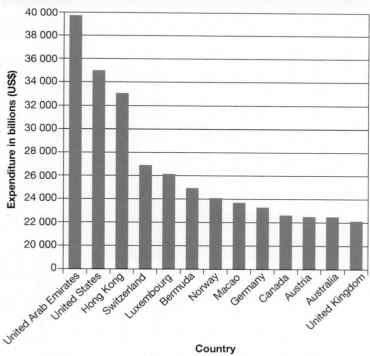

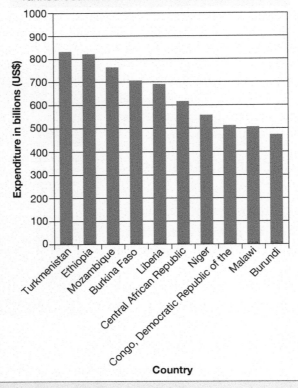

FIGURE 4 Household final consumption expenditure per person for the 10 lowest ranked countries in the world for 2013–2014

Expenditure in billions (US$)

Country

10.2 Activities

To answer questions online and to receive **immediate feedback** and **sample responses** for every question, go to your learnON title at www.jacplus.com.au. *Note*: Question numbers may vary slightly.

Remember

1. What reasons can you suggest as to why goods and services are traded?
2. Name the four levels of industry, and give an example of a good as it moves through the production process.

Explain

3. What reasons can you suggest for component parts of the Airbus A380 having to come from different *places* (countries)?
4. Explain what is meant by the term *value adding*, as a product moves through the four levels of industry. Choose a product such as wheat or timber.

Discover

5. Suggest why the United States is one of the largest consumers of goods and services in the world.
6. Use the **CIA** weblink in the Resources tab to select a country to compare with Australia. Look up figures for population and trade and see if you can explain why the figures for the country you have chosen differ from the figures for Australia, based on the wider range of data presented for each country.

Predict

7. It has been claimed that countries such as China and India, with growing middle classes that are now eager for goods and services, will put a strain on world resources. How might a growing demand for energy sources in these countries affect the *environment*?
8. How might a *change*, such as growth in Australia's population from 23 million to 40 million, affect Australia's trade?

Think

9. Why is Hong Kong, which is a Self Administered Region (SAR) of China, ranked third in household final consumption per person, while the wider country of China is ranked one-hundred-and-twelfth?
10. What reasons can you give for people being able to survive on $468 per person in countries such as Burundi?

10.3 How does trade connect Australia with the world?

10.3.1 Australia's trade organisation

Australia is one of the 162 members of the World Trade Organization (WTO), which covers 95 per cent of global trade. The organisation promotes free and fair trade between countries and, since 2001, its Doha Development Agenda has aimed to help the world's poor by slashing **trade barriers** such as tariffs, quotas and farm subsidies.

FIGURE 1 People involved with trade

The Australian Department of Foreign Affairs and Trade (DFAT) coordinates trade agreements on behalf of the Australian Government, and the Australian Trade Commission (Austrade) promotes the export of goods and services. About 73 per cent of Australia's trade is with the member countries of the Asia–Pacific Economic Cooperation (APEC) forum.

10.3.2 Australia's trading partners

China, Japan, the Republic of Korea and the United States were Australia's top four **trading partners** in 2014. Figure 2 and table 1 show details of Australia's import and export trade for the years 2013 and 2014.

FIGURE 2 Australia's top five import and export partners, 2013 (A$ million)

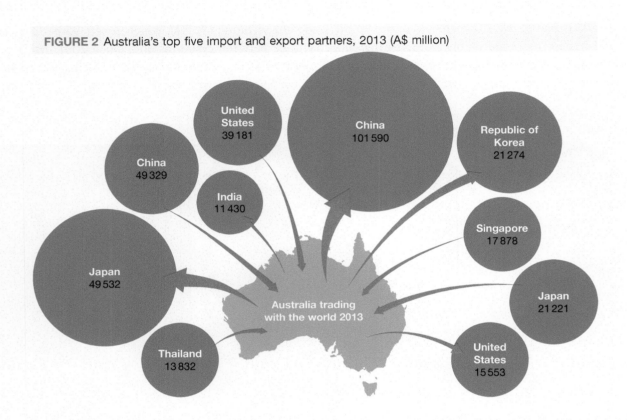

TABLE 1 Australia's top 10 two-way trading partners, 2014 (A$ million)

Rank		Goods	Services	Total	% share
	Total two-way trade	533 264	130 555	663 819	
1	China	142 078	10 390	152 468	23.0
2	Japan	65 431	4 787	70 218	10.6
3	United States	40 635	19 807	60 442	9.1
4	Republic of Korea	32 424	2 202	34 626	5.2
5	Singapore	21 128	9 059	30 187	4.5
6	New Zealand	16 083	7 384	23 467	3.5
7	United Kingdom	9 920	10 868	20 788	3.1
8	Malaysia	17 394	3 188	20 582	3.1
9	Thailand	16 106	2 872	18 978	2.9
10	Germany	13 725	3 000	16 725	2.5

10.3.3 Australia's types of trade

Exports

Australia's export trade in 2014 was valued at $327 billion, and was dominated by the mineral products of iron ore, coal and natural gas. Recreational and educational travel were Australia's leading services exports. See figure 3 for details of leading exports.

International students

A more recent high-level earner for Australia (now ranked as our number four export after iron ore, coal and natural gas), is the category of 'education-related travel services', which for 2015 was valued at $18.2 billion.

With more than 450 000 international students from 191 countries studying in Australia (see figure 4), education is now very important for our economy. In effect, it is a service export in that students are paying for knowledge that they will take back to their home country.

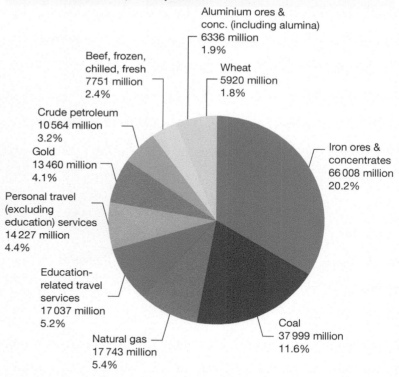

FIGURE 3 Australia's exports of goods and services for 2014 — share of exports by sector

- Aluminium ores & conc. (including alumina) 6336 million 1.9%
- Wheat 5920 million 1.8%
- Beef, frozen, chilled, fresh 7751 million 2.4%
- Crude petroleum 10 564 million 3.2%
- Gold 13 460 million 4.1%
- Personal travel (excluding education) services 14 227 million 4.4%
- Education-related travel services 17 037 million 5.2%
- Natural gas 17 743 million 5.4%
- Iron ores & concentrates 66 008 million 20.2%
- Coal 37 999 million 11.6%

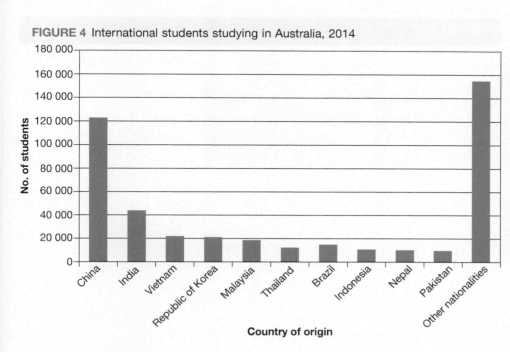

FIGURE 4 International students studying in Australia, 2014

(Bar chart: No. of students vs Country of origin)
China ~123 000, India ~44 000, Vietnam ~22 000, Republic of Korea ~21 000, Malaysia ~18 000, Thailand ~11 000, Brazil ~14 000, Indonesia ~10 000, Nepal ~9 000, Pakistan ~9 000, Other nationalities ~153 000

FIGURE 5 Education is now a major export for Australia, with the number of international students studying here increasing in recent years.

Imports

Like many countries, Australia is not self-sufficient in all goods and services. In 2014 Australia imported goods and services to a value of close to $337 billion. Figure 6 shows the main aspects of this trade.

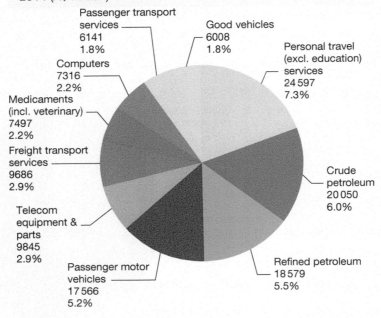

FIGURE 6 Australia's leading imports for goods and services, 2014 (A$ million)

Passenger transport services
6141
1.8%

Good vehicles
6008
1.8%

Personal travel (excl. education) services
24 597
7.3%

Computers
7316
2.2%

Medicaments (incl. veterinary)
7497
2.2%

Freight transport services
9686
2.9%

Telecom equipment & parts
9845
2.9%

Passenger motor vehicles
17 566
5.2%

Refined petroleum
18 579
5.5%

Crude petroleum
20 050
6.0%

FIGURE 7 Oil and petroleum products make up a significant part of Australia's import trade.

10.3 Activities

To answer questions online and to receive **immediate feedback** and **sample responses** for every question, go to your learnON title at www.jacplus.com.au. *Note*: Question numbers may vary slightly.

Remember

1. What is the *interconnection* between the World Trade Organization and Australia's trade?
2. What are Australia's three most important exports and imports?

Explain

3. Refer to figure 2 and table 1. Compare Australia's imports and exports with those of Asian countries. What changes have occurred in the imports and exports between countries from 2013 to 2014. Use data to support your answer.

4. Despite having a relatively small population, Australia has many goods and services to trade. Explain why this might be so.

Predict

5. How might a *change* in the growth of Australia's population affect the country's agricultural exports?
6. Look at the list of goods that Australia imports (see figure 6). What factors could lead to a *change* in the types of goods imported by the year 2050?

Think

7. (a) What evidence is there in this section to confirm the fact that Australia is regarded as mostly a **primary industry** exporter?
   (b) Are there any figures of export trade that contradict this statement?
   (c) Why has Australia become such an important exporter of education services?

Try out this interactivity: Trading partners (int-3338)

Deepen your understanding of this topic with related case studies and questions.
○ Aid, migration and trade

10.4 How is food traded around the world?

10.4.1 Trade in food

The world's population is unevenly distributed across space, as is the quantity of food produced. Some places, such as Australia, produce an abundance of food, while others struggle to produce enough to maintain food security.

Traditionally, food production consisted of hunting and gathering or cropping and herding. Excess food was consumed locally or sent to nearby markets for **barter** or cash. While up to 41 per cent of the world's population is still directly tied to subsistence agriculture, many of the world's highly developed economies produce large surpluses of food specifically for international trade. For instance, Australia's food production is estimated to be worth $100 billion annually, and of that, $44.3 billion-worth is exported.

The flow of food trade

Much of the flow of food trade is controlled by powerful nations, such as the United States, the European Union and China, and by international food trade agreements. The World Trade Organization (WTO) and G20 (a group of 20 developed and powerful nations) have a significant say in the flow of food products around the world, particularly with respect to tariffs and fair trade rules.

Food trade is a complicated business, as can be seen in figure 1. It is estimated that for **developing countries**, three-quarters of exports are agricultural produce. While developed countries may need to import tropical foods, many actually export as much as they import in agricultural produce. For instance, the United States, Canada and Australia use large farms to produce wheat, and they control 75 per cent of the global export trade in cereals. In 2015, Australia produced 23 million tonnes of wheat, the majority of which was for sale in overseas markets (see figure 2). Australia's top 10 agricultural markets can be seen in figure 3.

FIGURE 1 World trade flows — exports of agricultural products by region, US$ billion

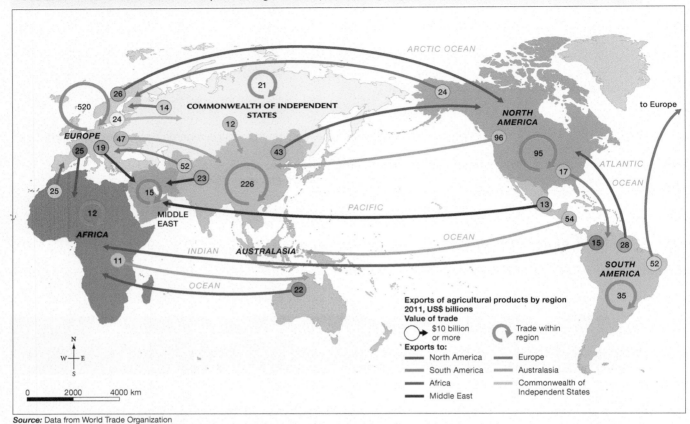

Source: Data from World Trade Organization

FIGURE 2 Top export countries for Australian wheat

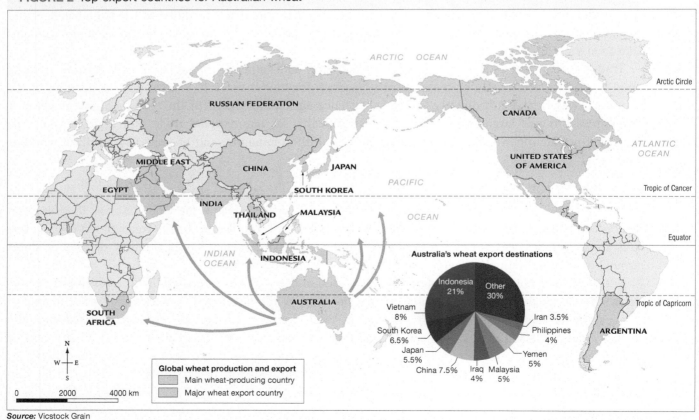

Source: Vicstock Grain

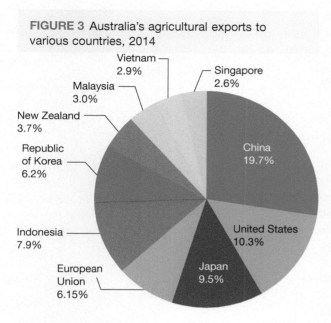

FIGURE 3 Australia's agricultural exports to various countries, 2014

- Vietnam 2.9%
- Singapore 2.6%
- Malaysia 3.0%
- New Zealand 3.7%
- Republic of Korea 6.2%
- China 19.7%
- Indonesia 7.9%
- United States 10.3%
- European Union 6.15%
- Japan 9.5%

10.4.2 Trade in animals

World trade in animals as food is estimated at close to 50 million animals per year — pigs, cattle, goats and sheep. Using modern shipping methods, many animals are transported over long distances, and questions have been asked about potential cruelty in the operation of this trade (see figure 4).

Australia's exports of live sheep and cattle to Asia and the Middle East in 2014 amounted to 3.7 per cent of Australia's total exports. The value of the live animal export trade to Australia is estimated at $1.6 billion or 3.7 per cent of all agricultural produce per year. The industry employs up to 10 000 rural workers at abattoirs, ports and in the transport industry. While there may be concerns about this industry, it should be remembered that countries request live animal exports so that they can be slaughtered according to **halal** religious customs.

Due to the extensive nature of farming cattle and sheep in Australia, these animals must often travel very large distances to reach ports. They then travel by ship to distant markets. The Australian Government has set high standards in the handling of live animals and is monitoring carefully how they are treated at destination ports.

FIGURE 4 Live animal transport around the world

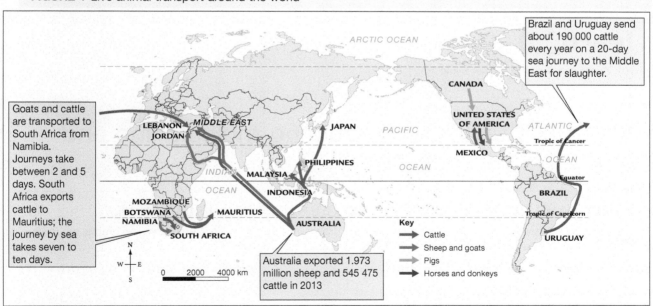

Brazil and Uruguay send about 190 000 cattle every year on a 20-day sea journey to the Middle East for slaughter.

Goats and cattle are transported to South Africa from Namibia. Journeys take between 2 and 5 days. South Africa exports cattle to Mauritius; the journey by sea takes seven to ten days.

Australia exported 1.973 million sheep and 545 475 cattle in 2013

Key
- Cattle
- Sheep and goats
- Pigs
- Horses and donkeys

0 2000 4000 km

Source: Spatial Vision

10.4 Activities

To answer questions online and to receive **immediate feedback** and **sample responses** for every question, go to your learnON title at www.jacplus.com.au. *Note*: Question numbers may vary slightly.

Remember

1. What is the value of Australia's agricultural exports?
2. What is the percentage of world trade in animals that Australia controls?
3. Refer to figure 1.
   (a) What is the value of food trade from Australasia to Europe?
   (b) What is the value of food trade from Europe to Australasia?
   (c) Is there a balance in this food trade based on your calculations in (a) and (b)?
   (d) Place the regions of the world in decreasing order by volume of food trade.

Explain

4. Refer to figure 2.
   (a) Explain why Australia can export such a large quantity of wheat to the world.
   (b) What reasons can you suggest as to why a country such as Russia might not export wheat to Indonesia and Malaysia?

Discover

5. Why do countries in *places* such as the Middle East and Asia have a preference for live animal imports?
6. Use the **Trade** weblink in the Resources tab to help you prepare a 200-word statement about the types of agricultural produce Australia exports and the nations to which we export. Also state which agricultural produce Australia imports.
7. (a) Investigate the issue of live animal exports from Australia.
   (b) How might a ban on live animal exports from Australia affect farmers?

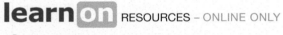 RESOURCES – ONLINE ONLY

 Explore more with this weblink: Trade

10.5 SkillBuilder: Constructing multiple line and cumulative line graphs

WHAT ARE MULTIPLE LINE GRAPHS AND CUMULATIVE LINE GRAPHS?

Multiple line graphs consist of a number of separate lines drawn on a single graph. Cumulative line graphs are more complex to read, because each set of data is added to the previous line graph.

Go online to access:

- a clear step-by-step explanation to help you master the skill
- a model of what you are aiming for
- a checklist of key aspects of the skill
- a series of questions to help you apply the skill and to check your understanding.

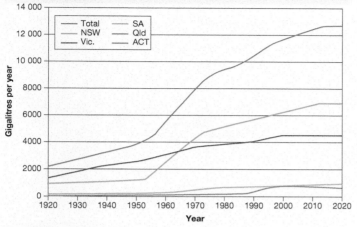

FIGURE 1 Water use by five states: a multiple line graph

Legend: Total, NSW, Vic., SA, Qld, ACT

y-axis: Gigalitres per year (0 to 14 000)
x-axis: Year (1920 to 2020)

Source: Food and Agriculture Organization of the United Nations, 2012 FAOSTAT, http://faostat3.fao.org/home/index.html

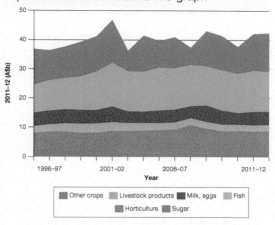

FIGURE 2 Value of farms and fisheries food production: a cumulative line graph

y-axis: 2011-12 (A$b) (0 to 50)
x-axis: Year (1996–97 to 2011–12)

Legend: Other crops, Livestock products, Milk, eggs, Fish, Horticulture, Sugar

Source: © DAFF 2013, *Australian Food Statistics 2011-12.* Department of Agriculture, Fisheries and Forestry, Canberra. CC BY 3.0.

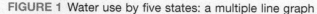

learn on RESOURCES – ONLINE ONLY

Watch this eLesson: Constructing multiple line and cumulative line graphs (eles-1740)

Try out this interactivity: Constructing multiple line and cumulative line graphs (int-3358)

10.6 How has the international automotive trade changed?

10.6.1 The car trade

Trade is one of the strongest interconnections between people and places on the planet, as goods are transported all over the world. Worldwide earnings from car exports make up 5.3 per cent of earnings from all international exports. Two countries with the largest share of the car export market are Japan in north-east Asia and Germany in Europe. Together, they have 61 per cent of the world's net profits on international car exports. Other countries with significant car exports include China and South Korea in Asia.

10.6.2 The rise of the Asian car industry

The Honda Motor Company is a Japanese multinational corporation. It is:
- the world's largest motorcycle manufacturer
- the world's largest manufacturer of internal combustion engines (by volume)
- the second largest Japanese car manufacturer
- the fourth largest car manufacturer in the United States
- the seventh largest car manufacturer in the world behind Toyota, General Motors, Volkswagen, Hyundai, Ford and Nissan.

The company has assembly plants around the globe (see figure 1). Honda develops vehicles to cater to different needs and markets, so its vehicles vary by country and may feature vehicles exclusive to a region.

FIGURE 1 The Honda Motor Company operates plants worldwide for the manufacture of its products.

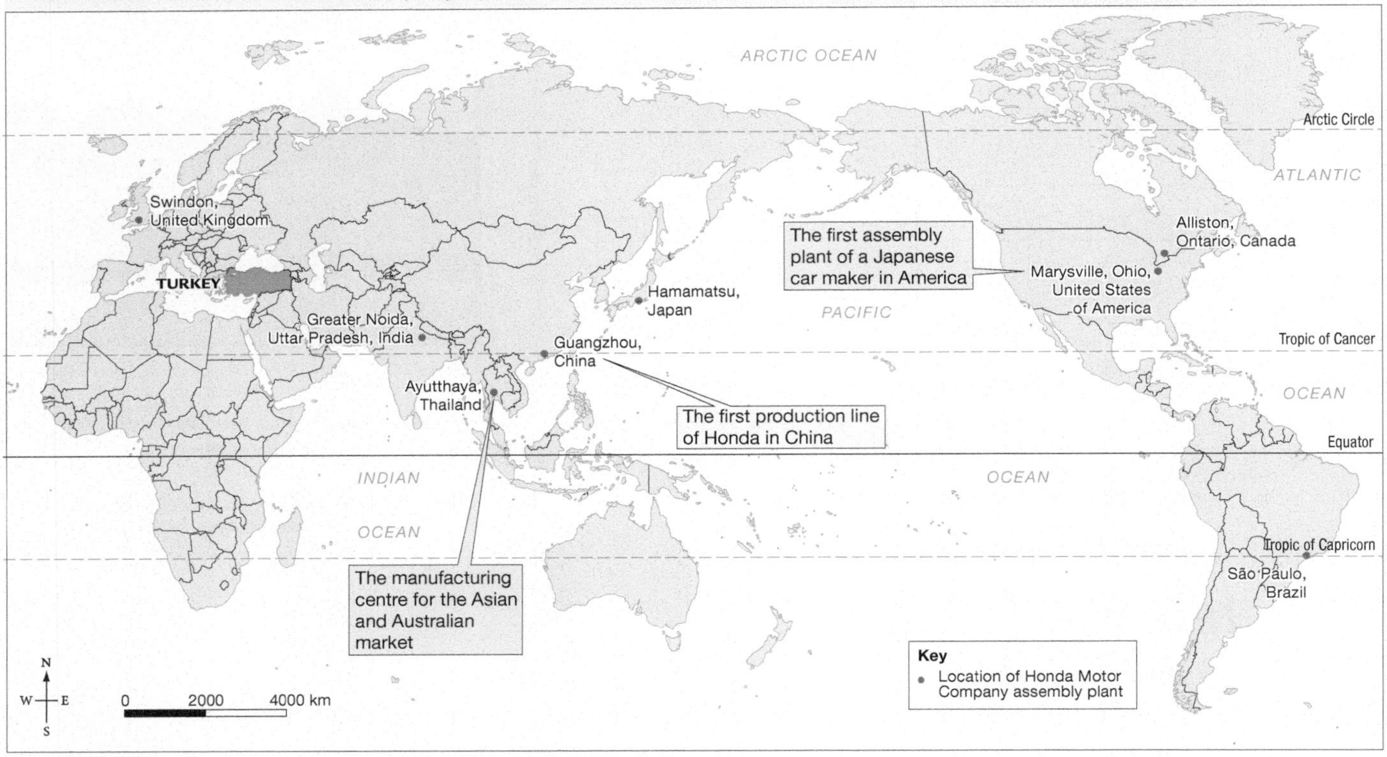

ARCTIC OCEAN

Arctic Circle

ATLANTIC

Swindon,
United Kingdom

Alliston,
Ontario, Canada

TURKEY

The first assembly
plant of a Japanese
car maker in America

Marysville, Ohio,
United States
of America

Hamamatsu,
Japan

PACIFIC

Tropic of Cancer

Greater Noida,
Uttar Pradesh, India

OCEAN

Guangzhou,
China

Ayutthaya,
Thailand

The first production line
of Honda in China

Equator

INDIAN

OCEAN

OCEAN

OCEAN

Tropic of Capricorn

São Paulo,
Brazil

The manufacturing
centre for the Asian
and Australian
market

Key
● Location of Honda Motor
Company assembly plant

N
W — E
S

0   2000   4000 km

Source: Spatial Vision

10.6.3 The decline of the American car industry

FIGURE 2 A Honda S2000

Detroit is the largest city in the state of Michigan, in the United States, and was long known as the world's automotive centre. Detroit and the surrounding region constitute a major centre of commerce and global trade, most notably as home to America's 'Big Three' automotive companies: General Motors, Ford and Chrysler.

These three car manufacturers were, for a while, the largest in the world, and two of them are still in the top five. Ford had held the position of second-ranked car maker for 56 years, but was relegated to third in North American sales after being overtaken by Toyota in 2007. That year, Toyota also produced more vehicles than General Motors. In the North American market, the Detroit car makers retain the top three spots, though their market share is dwindling, and Honda passed Chrysler for the fourth spot in the United States in 2008.

The United States car industry is suffering from increased overseas competition and from the 2008–2012 global recession. Car dealerships across the United States are struggling, and many are closing. The Big Three manufacturers have suffered from perceived inferior quality compared to their Japanese counterparts. They have also been slow to bring new vehicles to the market, whereas the Japanese are considered leaders in producing smaller, more fuel-efficient cars.

Falling sales and market share have resulted in the Big Three's plants operating below capacity. General Motors' plants were at 85 per cent capacity in November 2005 — well below the plants of its Asian competitors. This led to production cuts, plant closures and layoffs. Between 2000 and 2010, Detroit's population fell by 25 per cent. It had been the nation's tenth largest city but is now its eighteenth largest, with a population of 713777. In April 2012, General Motors and Ford continued to lose market share, with yearly sales down 8.2 per cent and 5.3 per cent respectively.

As a consequence, a rise in automated manufacturing using **robotic technology** has created related industries in Detroit. Inexpensive labour in other parts of the world and increased competition have led to a steady change in certain types of manufacturing jobs in the region. For example, the Detroit area has gained new **lithium ion battery** plants. As well, the Detroit car makers and local manufacturers have restructured in response to competition. General Motors has invested heavily in fuel-cell-equipped vehicles, while Chrysler has researched and developed **biodiesel**.

FIGURE 3 Car industry redistribution in the United States

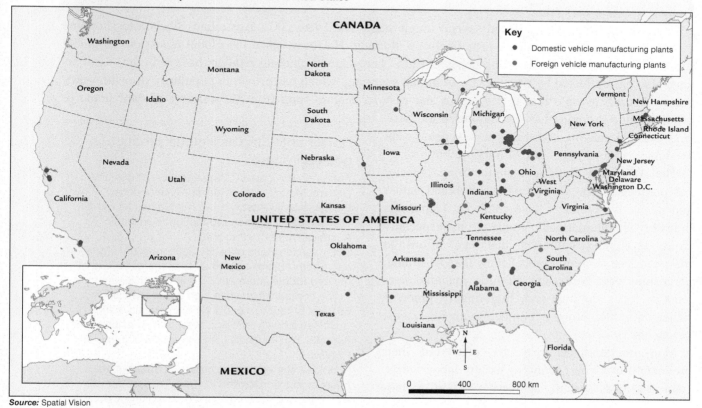

Source: Spatial Vision

10.6 Activities

To answer questions online and to receive **immediate feedback** and **sample responses** for every question, go to your learnON title at www.jacplus.com.au. *Note:* Question numbers may vary slightly.

Remember

1. Refer to figure 1. Honda cars are manufactured in many *places* around the world. Where is Honda's manufacturing centre for the Australian market?

Explain

2. Explain the advantages of the Big Three US car manufacturers being located in the same *place*.
3. Why has the Japanese automotive industry become a world leader in international exports?
4. What *changes* have Detroit's industries made in response to the decline of the car industry?
5. Look at figure 3. Describe the location of foreign and domestic automotive plants.

Discover

6. Research the effects of the decline of the US car industry on Detroit and surrounding towns and businesses.
7. New Zealand used to have a Honda assembly plant. Find out when it opened, when it closed and the reason for its closure.

10.7 Are global players altering the industrial landscape?

10.7.1 Changing trends

Until recently, designer clothing came from Italy, and jeans came from the United States. Today, that Italian suit might be designed in Milan, but it is woven from New Zealand wool and stitched together in China. **Globalisation** has brought global marketing, encouraging consumers everywhere to buy goods without a thought to where they come from.

The clothing industry has faced several tough years. Until late 2008, Australians flocked to shopping centres around the country and spent freely on clothing. However, since the global recession, or global financial crisis, consumer spending behaviour has changed, and clothing retailers have suffered. Retailers have also been affected by the increase in online shopping, which has grown in popularity over the years, mainly because people find it convenient and easy to bargain shop from the comfort of their home or office. Online shopping has revolutionised the business world by making everything anyone could want available by the simple click of a mouse button.

The Australian clothing manufacturing industry has produced some very recognisable brand names and distinctive products. One challenge facing this industry is international competition, especially from developing countries that can afford to mass-produce clothing more cheaply than Australian companies can. Australian clothing manufacturers tend to focus on high-end, high-quality products rather than attempting to compete with lower cost producers.

Many multinational firms have '**offshored**' their production to China, owing to its low labour costs, stable political system, aggressive export promotion policies and exchange rate. There are no labour unions, and incentives offered by the government include tax breaks, low import duties, low-cost land and low construction costs for new factories.

Horror unfolding as Holden production goes off-shore

After months of speculation and days of denials, General Motors Holden has finally announced the decision its Australian workforce has feared. The company will stop production of its iconic Australian car brand from 2017.

It means the loss of nearly 3000 jobs in Victoria and South Australia and may spell the beginning of the end for the automotive industry in this country.

The company blames a high dollar, the rising costs of production and a small domestic market.

Holden follows Ford and Mitsubishi in exiting Australia and it signals the demise of a cultural icon.

Holden started making cars in Australia in 1917. Mass production started in 1948. The last car will roll off the production line in 2017, making it a century of car making.

Source: ABC Online 11 December 2013

FIGURE 1 This symbol signifies that a product has been manufactured in Australia by an Australian-owned company.

FIGURE 2 Ikea's store in Beijing had more than six million visitors in 2011.

10.7.2 Foreign companies in China

In 1979, there were 100 foreign-owned enterprises in China. In 1998, there were 280 000, and by 2012 there were 436 800. Since 2007, foreign companies have employed 25 million people in China. Foreign companies that operate in China include Coca-Cola, Pepsi, Nike, Citibank, General Motors, Philips, Ikea, Microsoft and Samsung.

FIGURE 3 Top 30 locations for offshore companies in 2010, by region

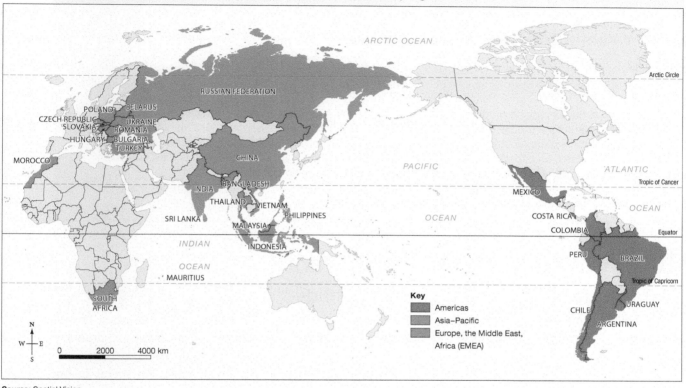

Source: Spatial Vision

10.7.3 Are your clothes made in sweatshops?

If you buy the brands Puma, Nike, Adidas, Mizuno, New Balance, Bonds or Just Jeans, then you may be wearing clothing or footwear that was made in a sweatshop.

A sweatshop is any working environment in which the workers experience long hours, low wages and poor working conditions. Typically, they are workshops that manufacture goods such as clothing. Sweatshops are common in developing countries, where labour laws are less strict or are not enforced at all. Workers have to use dangerous machinery in cramped conditions and can even be exposed to toxic substances. Child labour may be used.

Despite the fact that sweatshop workers receive wages, many of them continue to live in poverty. Most are young women aged 17 to 24.

FIGURE 4 A Bangladesh sweatshop

10.7 Activities

To answer questions online and to receive **immediate feedback** and **sample responses** for every question, go to your learnON title at www.jacplus.com.au. *Note*: Question numbers may vary slightly.

Remember

1. Why have many countries moved their production to offshore *places*?
2. What are sweatshops?

Predict

3. What *change* do you think online shopping will make to the Australian retail industry?
4. What do you think would happen to the price of clothing if sweatshops were to close down?

Think

5. Look at figure 3. Give reasons why most offshore manufacturing companies are located in the Asia–Pacific region.
6. What impact does moving production offshore have on the Australian economy and its people?
7. If clothing carries the Ethical Clothing Australia (ECA) label, it means the garment was manufactured in Australia and the manufacturer has ensured that all people involved in its production received the legally stated wage rates and conditions — known in Australia as award wages and conditions. Find out which Australian-made garments you can purchase to support fair working conditions.
8. How might internet shopping affect *places* such as shopping centres?
9. Are sweatshops ethical or *sustainable*? Explain your answer.

10.8 Why is fair trade important?

10.8.1 Social justice and problems of trade

The benefits of international trade are not evenly distributed around the world, and trade often favours large, developed countries rather than small developing countries. It is the role of governments, organisations and agencies to regulate this trade so that economic benefits are more evenly distributed.

Australians benefit economically, culturally and politically from international trade, but **social justice** problems can arise from this trade. For example, when we import blood diamonds from Africa or carpets made by child labour from Nepal, we are supporting unethical industries.

In addition, some countries can make it difficult for other countries to compete fairly, on a 'level playing field'. They do this by:
- imposing tariffs — taxes on imports
- imposing quotas — limits on the quantity of a good that can be imported
- providing subsidies — cash or tax benefits for local farmers or manufacturers.

10.8.2 Fair trade

The fair trade movement aims to improve the lives of small producers in developing nations by paying a fair price to artisans (craftspeople) and farmers who export goods such as handicrafts, coffee, cocoa, sugar, tea, bananas, cotton, wine and fruit. The movement operates through various national and international organisations such as the World Fair Trade Organization and Fairtrade International (see figure 2).

FIGURE 1 Fair trade organisations promote fair labour practices such as preventing and eliminating child labour.

FIGURE 2 Goods produced by workers for the World Fair Trade Organization mission

The fair trade labelling system is operated by Fairtrade International, of which Australia is a participating member. This system works to ensure that income from the sale of products goes back directly to farmers, artisans and their communities. Fairtrade International is present in over 120 countries worldwide and has a total of 1226 producer groups (see figure 3). The number of farmers and workers participating in Fairtrade is estimated to be 1.4 million, with 23 per cent of this number being women.

FIGURE 3 Fairtrade in the world, 2013

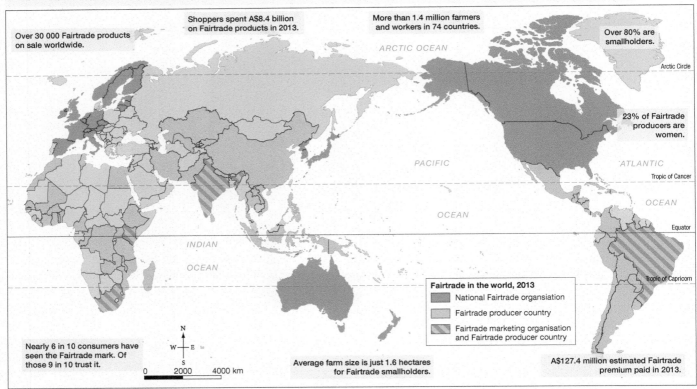

Over 30 000 Fairtrade products on sale worldwide.

Shoppers spent A$8.4 billion on Fairtrade products in 2013.

More than 1.4 million farmers and workers in 74 countries.

Over 80% are smallholders.

23% of Fairtrade producers are women.

Nearly 6 in 10 consumers have seen the Fairtrade mark. Of those 9 in 10 trust it.

Average farm size is just 1.6 hectares for Fairtrade smallholders.

A$127.4 million estimated Fairtrade premium paid in 2013.

Fairtrade in the world, 2013
- National Fairtrade organsiation
- Fairtrade producer country
- Fairtrade marketing organisation and Fairtrade producer country

0   2000   4000 km

Source: Fairtrade Foundation

Fairtrade food items currently include sugar, chocolate, coffee, tea, wine and rice. Other products include soaps, candles, clothing, jewellery, bags, rugs, carpets, ceramics, wooden handicrafts, toys and beauty products.

In 2013, Australia and New Zealand had a combined retail sales total of A$290.3 million in Fairtrade Certified products. The highest volume of trade in tonnes for 2013 were bananas: 373 000; sugar: (cane sugar) 194 000; and coffee/cocao beans: 138 500. On a global scale, 11.4 million producers and their families have benefited from Fairtrade-funded infrastructure and community development projects with a value of $8.4 billion.

10.8.3 Non-government organisations and fair trade

Non-government organisations (NGOs) such as Oxfam and World Vision also support fair trade, and oppose socially unjust trade agreements. They oppose attempts by developed countries to:
- block agricultural imports from developing countries
- subsidise their own farmers while demanding that poor countries keep their agricultural markets open.

10.8 Activities

To answer questions online and to receive **immediate feedback** and **sample responses** for every question, go to your learnON title at www.jacplus.com.au. *Note:* Question numbers may vary slightly.

Remember

1. What are the main principles of fair trade?

10.9 Why does Australia give foreign aid?

10.9.1 Introduction

Countries give aid to other countries in order to help with their development and, for example, help people overcome poverty and resolve humanitarian issues. Aid often helps the donor country by promoting stability and prosperity in the region. Australia's Official Development Assistance (ODA) program is known as Australian Aid.

10.9.2 Australian Aid aims

Over one billion people in the world live in poverty and do not have easy access to education and health care. When disasters strike, they lack the resources to get back on their feet. Poverty needs to be addressed by the international community because it can:

- breed instability and **extremism**
- cause people to flee violence and hardship, thus swelling the number of refugees.

Australia takes the stance that helping people who are less fortunate is a vital way of supporting **humanitarian principles** and social justice. Apart from showing we care, it is in the interests of our **national security**. It also increases our status throughout the world and creates political and economic interconnections with our Asia–Pacific neighbours.

10.9.3 The Australian Aid program

The Department of Foreign Affairs and Trade (DFAT) manages the Australian Government's multi-billion-dollar overseas aid program. To ensure that funds reach the needy, Australian Aid works with Australian businesses, non-government organisations such as CARE Australia, and international agencies such as the United Nations (UN) and the World Bank.

10.9.4 The Australian Aid plan

Overseas aid is the transfer of money, food and services from developed countries such as Australia to developing countries such as Papua New Guinea. There are many programs that are part of the Australian Aid budget, which is classified as Official Development Assistance (ODA) (see figure 1). These include:

- aid to governments for post-war reconstruction, as in Afghanistan and Iraq

- distribution of food through the United Nations World Food Programme
- contributions to United Nations projects on refugees and climate change
- disaster and conflict relief in the form of food, medicine and shelter, such as that given after the 2004 Indian Ocean tsunami
- programs by non-government organisations to reduce child labour in developing countries
- funding for education programs
- support for Australian volunteers working overseas.

A newly developed Australian Aid program for 2015–16 includes $50 million for gender equity and women's economic empowerment.

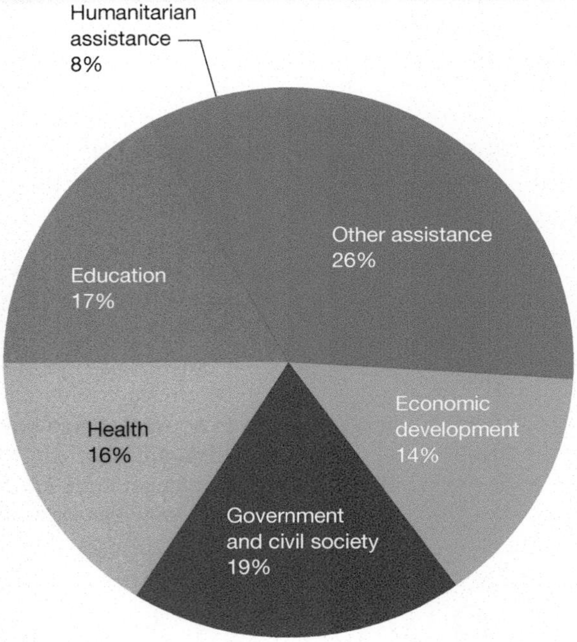

10.9.5 The Australian Aid budget

The Australian Aid program supports the United Nations Sustainable Development Goals (SDGs). In addition, countries with a low Human Development Index (HDI) are the target for development assistance. The Human Development Index ranks a country according to life expectancy at birth, adult literacy, school enrolments and income. The highest possible score for a country is 1.0 (see figure 2).

The total Australian Aid budget for 2015–16 was over 4 billion dollars with 92 per cent of the Regional Budget of 2.5 billion allocated to our nearest neighbours in the Asia–Pacific region (see figure 3).

FIGURE 1 Distribution of the Australian Aid budget

Humanitarian assistance — 8%

Other assistance 26%

Education 17%

Economic development 14%

Health 16%

Government and civil society 19%

FIGURE 2 The Human Development Index, 2015

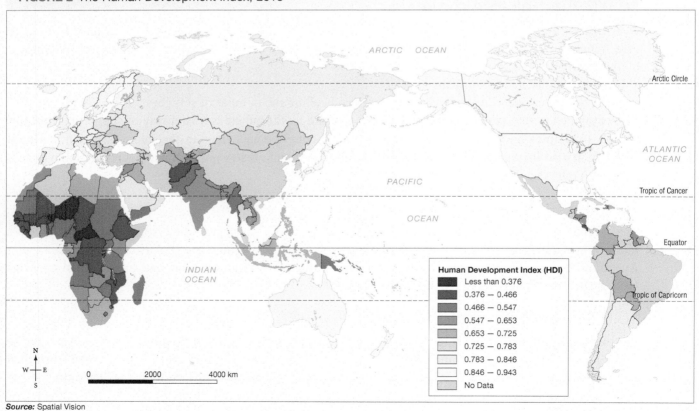

Human Development Index (HDI)
- Less than 0.376
- 0.376 — 0.466
- 0.466 — 0.547
- 0.547 — 0.653
- 0.653 — 0.725
- 0.725 — 0.783
- 0.783 — 0.846
- 0.846 — 0.943
- No Data

Source: Spatial Vision

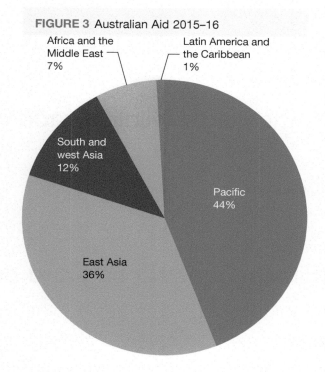

FIGURE 3 Australian Aid 2015–16

Africa and the Middle East 7%

Latin America and the Caribbean 1%

South and west Asia 12%

Pacific 44%

East Asia 36%

10.9 Activities

To answer questions online and to receive **immediate feedback** and **sample responses** for every question, go to your learnON title at www.jacplus.com.au. *Note*: Question numbers may vary slightly.

Remember

1. Which government department manages Australia's Official Development Assistance program?
2. Which regions of the world receive most of Australia's aid funding and why do you think this is so?

Explain

3. What reasons can you put forward to explain why Australian Aid programs are worthwhile in terms of Australia's *interconnections* with its neighbours?
4. Is there a case that could be argued for cutting aid budgets? Explain your reasons.

Discover

5. Undertake some internet research to find out how the Sustainable Development Goals guide the Australian Aid program.
6. (a) Why is East Asia such an important *place* for the distribution of Australian Aid?
   (b) Why does north-east Asia receive little foreign aid from Australia?

Predict

7. If Australian Aid were to stop, what *changes* do you think this would have on Australia's reputation in the international community?
8. Which elements of the Australian Aid program do you think will have the greatest impact on the lives of people in the Pacific region? Give reasons for your selection.

Think

9. Can you suggest a better way of distributing the budget dollars of the Australian Aid program in order to improve the lives of people in the Asia–Pacific region?
10. If Australia's economic prosperity were to decline in the next 50 years, which elements of the Australian Aid program do you believe would not be *sustainable*?

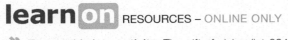
10.10 Why is the illegal wildlife trade a cause for concern?

Access this subtopic at **www.jacplus.com.au**

10.11 SkillBuilder: Constructing and describing a flow map

WHAT IS A FLOW MAP?

A flow map is a map that shows the movement of people or objects from one place to another. Arrows are drawn from the point of origin to the destination. Sometimes these lines are scaled to indicate how much of the feature is moving. Thicker lines show a larger amount; thinner lines show a smaller amount.

Go online to access:

- a clear step-by-step explanation to help you master the skill
- a model of what you are aiming for
- a checklist of key aspects of the skill
- a series of questions to help you apply the skill and to check your understanding.

FIGURE 1 Interstate migration flows, 1996 to 2001

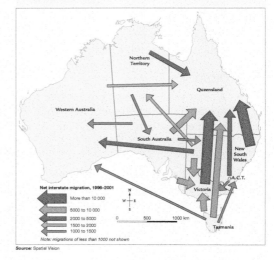

10.12 Review

10.12.1 Review

The Review section contains a range of different questions and activities to help you revise and recall what you have learned, especially prior to a topic test.

10.12.2 Reflect

The Reflect section provides you with an opportunity to apply and extend your learning.

Access this subtopic at **www.jacplus.com.au**

TOPIC 11
For better or worse?

11.1 Overview

Numerous **videos** and **interactivities** are embedded just where you need them, at the point of learning, in your learnON title at www.jacplus.com.au. They will help you to learn the content and concepts covered in this topic.

11.1.1 Introduction

The development of technology and communication around the world has had both positive and negative effects. One positive outcome is that advances in technology have made people's lives more interconnected. On the other hand, drawbacks include a widening gap between the 'haves' and the 'have-nots', and some negative effects on the environment. Therefore, it is essential to ask whether we are better off or worse off with ever-improving technology and communication.

E-waste at Agbogbloshie dump, in Accra, Ghana — a dumping ground for electronic waste from all over the developed world

Starter questions

1. What percentage of students in your class have a mobile phone? How many times a day do you use your mobile phone?
2. If you wanted to discard your outdated mobile phone, do you know how to do so responsibly? Do you think about where your e-waste goes?
3. Have you ever considered how many people in developing countries have access to the internet?
4. When did you last 'Google' the answer to a question?

INQUIRY SEQUENCE

11.1 Overview   208
11.2 How do you communicate?   209
11.3 Who has access to technology?   212
11.4 What are the consequences of unequal access?   214
11.5 How has technology improved lives in developing countries?   217
11.6 What are the impacts of e-waste production and consumption in China?   219
11.7 How are e-wastes managed?   223
11.8 **SkillBuilder:** Constructing a table of data for a GIS   online only   227
11.9 How does e-cycling work?   online only   228
11.10 How can you reduce your consumption?   228
11.11 **SkillBuilder:** Using advanced survey techniques — interviews   online only   230
11.12 **Review**   online only   231

11.2 How do you communicate?

11.2.1 Communication technology

Youth culture is strongly linked to the use of communication technology. More than 800 million adults in the world lack basic literacy skills, yet the boom in information and communication technology is skyrocketing. A survey conducted in 23 countries explored the media access and media use of 12-year-olds. The study showed that in 97 per cent of the countries surveyed, the inhabitants received at least one television channel. However, less accessible were personal computers (23 per cent) and the internet (9 per cent). Internet use among youths in developed countries is increasing at a rapid rate.

11.2.2 Who uses the internet?

Internet access in Australia is continually growing. Households with children have greater access to a computer and the internet at home than do households without children. This reflects the growing digital culture among youth (see figure 1).

It is often thought that internet communication is all done via satellite. However, that is not the case. There are also fibre optic cables under the sea that help link the network of networks that is the internet, as shown in figure 2. These cables improve the speed of the internet in Australia. New cable, which connects Australia to the United States via New Zealand, will assist in creating the national broadband network.

FIGURE 1 Households with access to the internet in Australia, 1998–2008

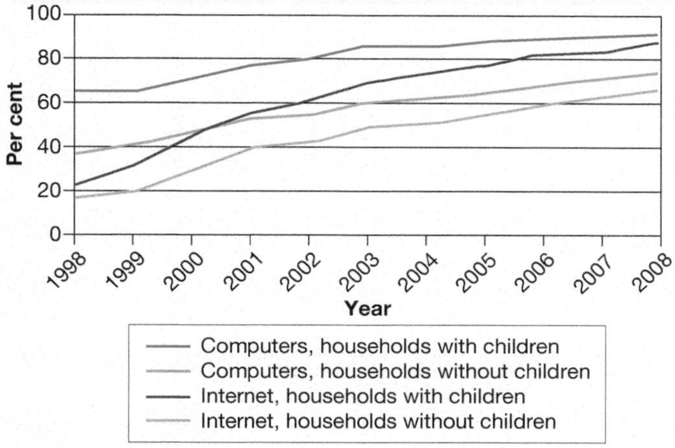

Legend:
- Computers, households with children
- Computers, households without children
- Internet, households with children
- Internet, households without children

Source: ABS 2008

FIGURE 2 The global submarine cable network, which helps connect the world's internet network

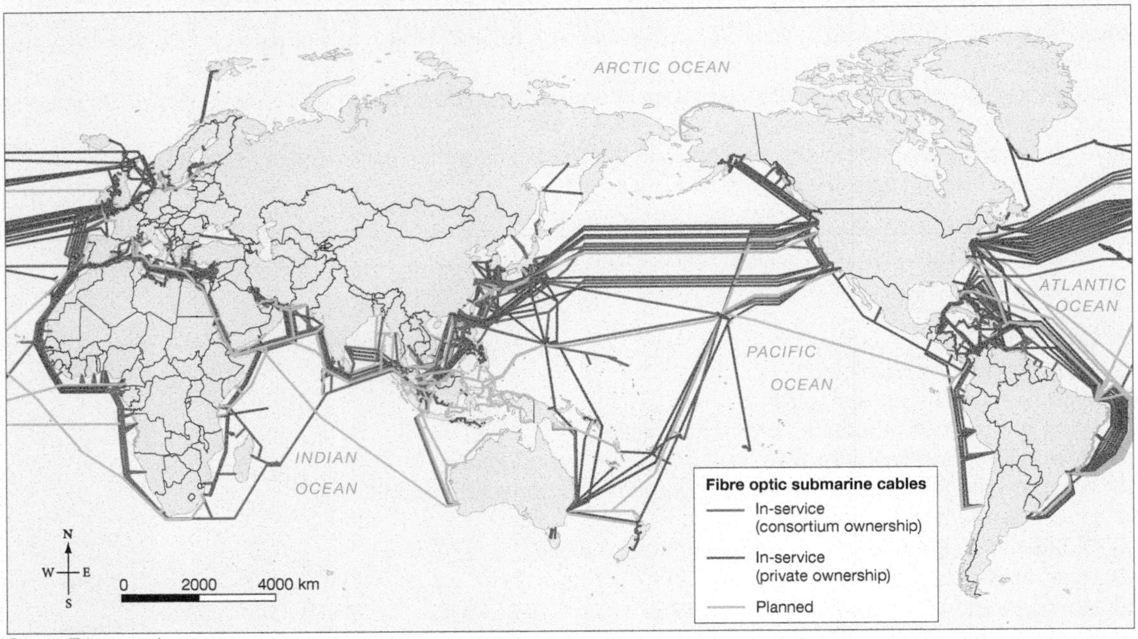

ARCTIC OCEAN

ATLANTIC OCEAN

PACIFIC OCEAN

INDIAN OCEAN

Fibre optic submarine cables
— In-service (consortium ownership)
— In-service (private ownership)
— Planned

N
W E
S
0 2000 4000 km

Source: Telegeography

11.2.3 What is Web 2.0?

The World Wide Web is a way of accessing and spreading information on the internet. **Web 2.0** technology enables the sharing of online information and ideas that anyone has created. It is the second generation of the World Wide Web, and enables collaboration and exchanging of information online. This is different from the original **Web 1.0**, which is considered to be read only (see figure 3). Web 1.0 allowed users only to search and read information. The nature of Web 2.0 makes it popular, because it is an easy form of communication technology to use. There is a variety of different types of Web 2.0 applications, including Facebook, YouTube, Wikipedia, Flickr and Instagram.

FIGURE 3 Web 1.0 versus Web 2.0

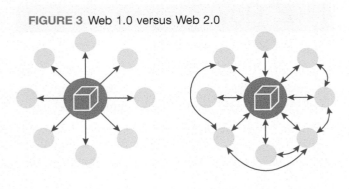

11.2.4 Mobile phones

Mobile phones are a form of wireless communication that is transmitted via radio wave or satellite transmission. They are no longer used solely for conversation. With new technology, mobile phones and smartphones are now a source of almost endless information. They can help people find their way from place to place, can be used as a diary, take photos and videos, print documents wirelessly, do banking, record music and much more. Mobile phones bridge the domains of communication and information technology. The Australian Bureau of Statistics (ABS) estimates that 29 per cent of all children owned a mobile phone in 2012. It was also noted that ownership increased with age. Two per cent of 5- to 8-year-olds owned mobile phones, while 73 per cent of 12- to 14-year-olds owned them.

11.2.5 Information and communication technology

Information and communication technology is now used for a variety of purposes to make connections. We often do not realise how convenient our lives have been made by various forms of communication.

Some of these include online and telephone banking, storing and sharing of medical records, online shopping and online games.

Skype is another example of an online communication medium — this service makes it possible for anyone with an internet connection to converse in the form of a video call with someone else anywhere in the world. Skype is also available as a mobile phone application, allowing any user to have a more interactive telecommunication experience. This simple form of communication technology improves the interconnectedness of places around the globe.

FIGURE 4 Skype allows you to make video calls using the internet.

11.2 Activities

To answer questions online and to receive **immediate feedback** and **sample responses** for every question, go to your learnON title at www.jacplus.com.au. *Note*: Question numbers may vary slightly.

Remember

1. Refer to figure 1. How many more households with children have an internet connection compared with households without children?
2. Refer to figure 2. How does the number of submarine cables in Australia compare with the number in Japan?

Explain

3. Refer to figure 2. Explain the geographical pattern of submarine cables worldwide. Where are the greatest concentrations of cables? Which parts of the seas or oceans seem to lack cables? Why do you think this is?

Discover

4. Refer to figure 3. Find out more about how web 1.0 differs from web 2.0, and create a PowerPoint presentation to compare and contrast them.

11.3 Who has access to technology?

11.3.1 Who owns mobile phones?

When we think of using the internet and mobile phones, we often forget about those who do not have access. Access is not equal across the world, let alone Australia. This is also related to government **expenditure** on the infrastructure needed to access information and communication technology.

Today it seems that mobile phones have become the most important form of communication. However, it is worth noting that not everyone in Australia owns one. As at June 2015, there were 21 million mobile phone users in Australia (out of a total population of 23.79 million). Most adults own at least one mobile phone, and many children also own phones (see figure 1).

Mobile phone use across the globe is also uneven. This could be due to a variety of factors, such as access, financial situation and way of life. Some countries, such as Saudi Arabia, have 180 mobile phones per 100 individuals, indicating that some people own two devices. In Bangladesh 76 mobile phones are owned per 100 individuals.

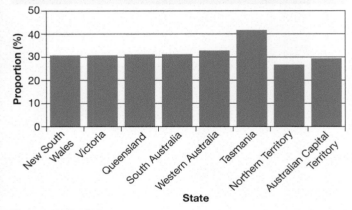

FIGURE 1 Mobile phone ownership among children aged 5–14 by Australian states and territories, 2009

Source: ABS 2009

11.3.2 Internet access

Some of us take for granted the fact that we can access the internet almost everywhere we go, whether we are at home, at school, at the shops, or even walking down the street. While this might not be the case for everyone in Australia, it is definitely not the case for everyone in the world (see figure 2).

11.3.3 ICT expenditure

It is important to note that the amount of money spent by individual countries on information and communication infrastructure can play a vital role in consumer access and uptake. Figure 3 indicates expenditure on information and communication technology (ICT) as a percentage of GDP. This map shows that expenditure across the globe is uneven and there is no set pattern. It is difficult to correlate access to ICT with country expenditure; for example, Australia spends between 3 and 6 per cent of its GDP, but access is relatively high. In contrast, a developing nation such as Bangladesh spends 9 to 12 per cent of its GDP, yet access is limited.

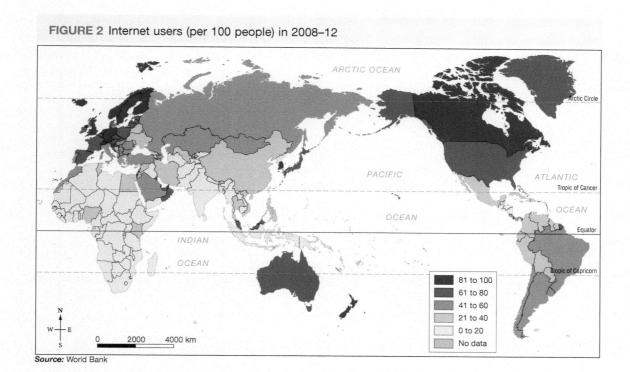

FIGURE 2 Internet users (per 100 people) in 2008–12

Legend:
- 81 to 100
- 61 to 80
- 41 to 60
- 21 to 40
- 0 to 20
- No data

Source: World Bank

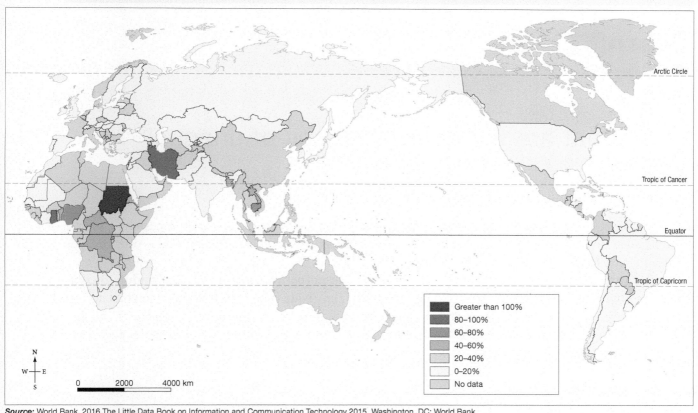

FIGURE 3 Expenditure on information and communication technologies by country, 2015

Legend:
- Greater than 100%
- 80–100%
- 60–80%
- 40–60%
- 20–40%
- 0–20%
- No data

Source: World Bank. 2016 The Little Data Book on Information and Communication Technology 2015. Washington, DC: World Bank

It is interesting to note that Estonia, a small country in Europe, sets a worldwide benchmark by providing free wireless internet access almost everywhere. A group of volunteers successfully lobbied Estonian cafés, hotels, hospitals, local governments and small business owners to provide this free access, which has been made available in a country with a GDP per capita of about US$20 000; considerably lower than that of Australia, which has a GDP per capita of about US$61 000. As a result, Estonian society has become reliant almost entirely on information in electronic form — doctors issue only electronic prescriptions, voting is conducted online, and government cabinet meetings are now paperless. Not only has free wifi helped the environment, it has also led to a better educated society.

11.3 Activities

To answer questions online and to receive **immediate feedback** and **sample responses** for every question, go to your learnON title at www.jacplus.com.au. *Note:* Question numbers may vary slightly.

Remember

1. Refer to figure 1. Which two states have the same percentage of mobile phone ownership?

Discover

2. Research the reasons for Sudan having a high expenditure on ICT.
3. Using figure 2 and the **Mobile phone subscriptions** weblink in the Resources tab, explain how global internet users and mobile phone subscribers are spatially *interconnected*. What are the similarities and differences?

Think

4. Refer to figure 2. What factors do you think affect the pattern of internet use throughout the world?

Predict

5. Refer to figure 3. Where in the world do you predict there will be an increase in infrastructure expenditure in the next 10 years? Explain.

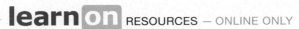

learn on RESOURCES — ONLINE ONLY

Explore more with this weblink: Mobile phone subscriptions

11.4 What are the consequences of unequal access?

11.4.1 What is the digital divide?

Unequal access to information technology creates a division between people with access and people without. Such divisions can have implications for economic growth and social equity. How often do you consider access to the internet as a necessity, not a luxury item?

There is a growth in internet access worldwide. However, the gap between low-income and **middle-income countries** is widening. This is known as the **digital divide**. The divide is primarily based on internet access, but it includes all forms of information and communication technology, as can be seen in tables 1 and 2.

It is also evident within countries that some people have access to high-speed internet, while others only have access to dial-up internet. This may be because of:
- income levels
- availability in local areas
- capabilities of computers and laptops
- the speed of internet access
- the level of technology assistance
- the price of connections
- capabilities of mobile phones.

TABLE 1 Middle-income countries, 2005 and 2014

	2005	2014
Sector performance		
Access		
Fixed-telephone subscriptions (per 100 people)	13.5	10.2
Mobile/cellular telephone subscriptions (per 100 people)	23.4	93.6
Fixed (wired) broadband subscriptions (per 100 people)	1.1	4.3
Households with a computer (%)	14.1	32.2
Households with internet access at home (%)	6.7	32.5
Usage		
International voice traffic, total (minutes/subscription/month)	3.8	4.6
Individuals using the internet (%)	7.1	34.1
Quality		
Population covered by a mobile/cellular network (%)	86	97

TABLE 2 Low-income countries, 2005 and 2014

	2005	2014
Sector performance		
Access		
Fixed-telephone subscriptions (per 100 people)	0.8	0.9
Mobile/cellular telephone subscriptions (per 100 people)	3.9	57.2
Fixed (wired) broadband subscriptions (per 100 people)	0.00	0.2
Households with a computer (%)	1.3	4.5
Households with internet access at home (%)	0.3	4.2
Usage		
International voice traffic, total (minutes/subscription/month)	—	—
Individuals using the internet (%)	1.0	6.3
Quality		
Population covered by a mobile/cellular network (%)	38	—

This gap does not necessarily divide society in two. It is also important to note that large populations do not always correlate with increased internet access (see figure 1).

FIGURE 1 Internet users (per 100 people)

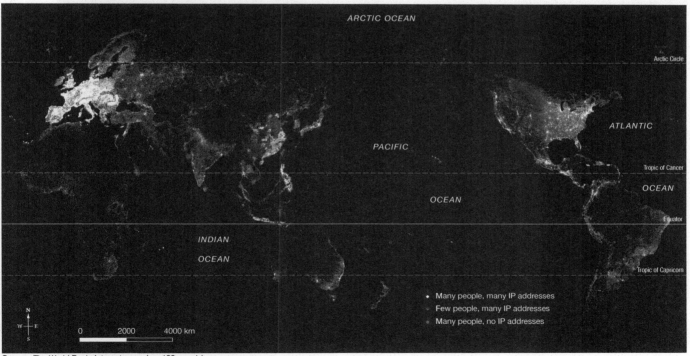

Source: The World Bank: Internet users (per 100 people)

The digital divide is not fixed or static; it changes as the structure and characteristics of the population shift. Digital disadvantage is a dimension of poverty, which leaves people with a sense of powerlessness. If your ability to connect to information and communication technology is limited, this can affect what you are able to do with your life.

11.4.2 Why bridging the gap is important

- *Economic equality*. Some social welfare services are administered electronically, so access to a telephone and the internet is important. The telephone also provides security, and can be used in emergency situations. In addition, access to the internet can be important for career development and accessing civic information.
- *Social mobility*. Computers play an important role in learning and education. The digital divide is unfair for children in lower socio-economic groups. The higher the qualification held by an individual, the more likely they are to have internet access at home.
- *Democracy*. Some people think that access to the internet creates a healthier democracy. It is thought that it increases public participation in elections and decision-making processes.
- *Economic growth*. Information and communication infrastructure and active use could stimulate economic growth, particularly for less developed nations. Economic changes and improvements tend to be associated with information technology. Certain industries can give a country's economy a competitive advantage.

11.4 Activities

To answer questions online and to receive **immediate feedback** and **sample responses** for every question, go to your learnON title at www.jacplus.com.au. *Note*: Question numbers may vary slightly.

Remember

1. Define the term *digital divide* in your own words.

Explain

2. Refer to tables 1 and 2.
   (a) Comparing low-income and middle-income countries, what *changes* have occurred between 2005 and 2014 to the percentage of individuals using the internet? Why?
   (b) Comparing low-income and middle-income countries, what *changes* have occurred between 2005 and 2014 to the percentage of fixed telephone subscriptions? Why?

Predict

3. Based on table 2, predict how many people in low-income countries will have access to mobile networks in 2100. Give reasons for your answer.

Think

4. Why do you think the uptake of mobile/cellular telephone subscriptions in middle-income countries has been significantly higher than in low-income countries? Explain.

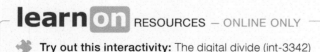 RESOURCES — ONLINE ONLY

Try out this interactivity: The digital divide (int-3342)

11.5 How has technology improved lives in developing countries?

11.5.1 The situation in Kenya

While there is a difference between the developed and developing world, it is important to note that access to the internet and mobile phone networks is improving. This is largely thanks to non-government agencies and to service providers. The people of Kenya, for example, no longer need to take long journeys into town or to wait in long queues just to transfer money. Instead, they can type in a couple of numbers, hit a button, and pay for anything they want within seconds via mobile phone.

Most people in Kenya live in rural and remote places in the countryside. Landline access is very limited, and it is also considered costly to install, owing to the vast distances. This can leave families disconnected from one another, as the primary earner in the family often has to work in a distant township in order to provide money for the family. Those left in the rural villages are often self-employed farmers or tradespeople. Without communication, families are often disconnected.

Improving access

Mobile phone coverage and access has been significantly improved in Kenya, which has dramatically improved people's lives. The UK-based non-government organisation (NGO) called Financial Deepening Challenge Fund (FDCF) has worked with Vodafone to set up M-Pesa, meaning 'mobile money'. A customer can go to an M-Pesa agent, such as a supermarket, and:

- deposit and withdraw money
- transfer money to other users and non-users
- pay bills
- purchase phone credit
- transfer money between the service and a bank account (see figure 1).

FIGURE 1 A customer at an M-Pesa agent

People do not need a bank account or even a permanent address to use M-Pesa. They receive a text message to confirm a transaction, and the money can then be stored on the phone or it can be forwarded to someone else. At any time, the money can be transferred back to cash through an M-Pesa agent or an ATM. Figures 2 and 3 show population and mobile phone coverage, indicating the coverage is significant in highly populated areas.

11.5.2 How successful has it been?

Today there are about 19 million subscribers and over 2200 M-Pesa agents in Kenya. These agents include petrol stations, supermarkets and other retail outlets. By 2010, over 50 per cent of Kenya's population was using M-Pesa. In 2012, when M-Pesa celebrated its fifth birthday, it had recorded over 15 million transactions. By March 2014 there were 260 transactions taking place every second. Figure 4 reflects this change in mobile phone usage and mobile money customers.

Taxi drivers have benefited, too. They no longer need to drive around with a lot of cash in their cars, because their passengers are able to pay using M-Pesa.

FIGURE 2 Kenya, population density

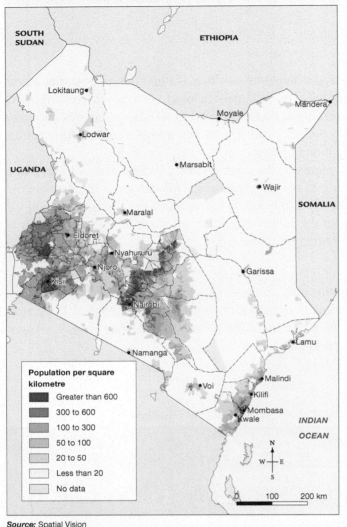

Source: Spatial Vision

FIGURE 3 Kenya, mobile phone coverage

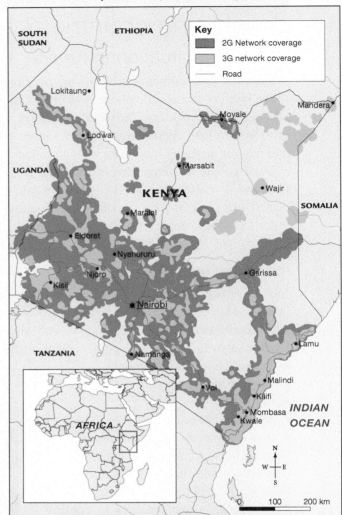

Source: A map of mobile coverage, Omae Malack Oteri, Langat Philip Kibet and Ndung'u Edward N., "Mobile Subscription, Penetration and Coverage Trends in Kenya's Telecommunication Sector" International Journal of Advanced Research in Artificial Intelligence (IJARAI), 4(1), 2015. http://dx.doi.org/10.14569/IJARAI.2015.040101

It has also been seen as an improvement in personal safety, because it is a secure and easy way to transfer money.

Access to mobile phones for small business operators has meant they are now able to advertise to a larger audience, and are no longer dependent on word-of-mouth advertising. Clients can now contact business operators with ease. For those working away from home, it is a safe and easy way to send money back to families in the countryside. This has enabled Masai herdsmen to go to the markets and make purchases by using their mobile phones. M-Pesa has eliminated the need to carry large sums of cash to markets, and has reduced the number of thefts.

FIGURE 4 Changes to mobile phone subscriptions in Kenya

M-Pesa demonstrates how dreaming big but thinking locally can have a significant effect on the economic and social structure of a place, just through the use of a mobile phone.

11.5 Activities

To answer questions online and to receive **immediate feedback** and **sample responses** for every question, go to your learnON title at www.jacplus.com.au. *Note*: Question numbers may vary slightly.

Remember

1. How successful has M-Pesa been in Kenya? Ensure you use specific examples in your answer.
2. What *interconnections* can now occur between people because of M-Pesa?

Explain

3. Why would safety be a concern to people in Kenya? How does safety relate to information and communication technology in Kenya?
4. Which *places* in Kenya are well serviced by the mobile phone network?
5. Which *places* are not well serviced by the mobile phone network? Suggest reasons for this.

Discover

6. Using the internet, research how another developing country has improved its information and communication technology. Create a multi-modal presentation on your findings.

Think

7. Which services in Australia serve a similar purpose to M-Pesa? Explain what they are and how they operate.

11.6 What are the impacts of e-waste production and consumption in China?

11.6.1 Introduction

China is one of the largest producers and consumers of electronics. Almost 50 per cent of the major home appliances manufactured in China are exported. However, due to China's growing economy and the increase in its development, the other 50 per cent is attributed to local consumption. Figure 1 shows the changing nature of the sale of home appliances in China, which reflects growth in the national economy.

As a result, China has a growing **e-waste** problem from both domestic and international markets. The informal processing of e-waste has resulted in damage to local environments and people's health. Until formal policies are established, informal recycling will co-exist with formal recycling companies.

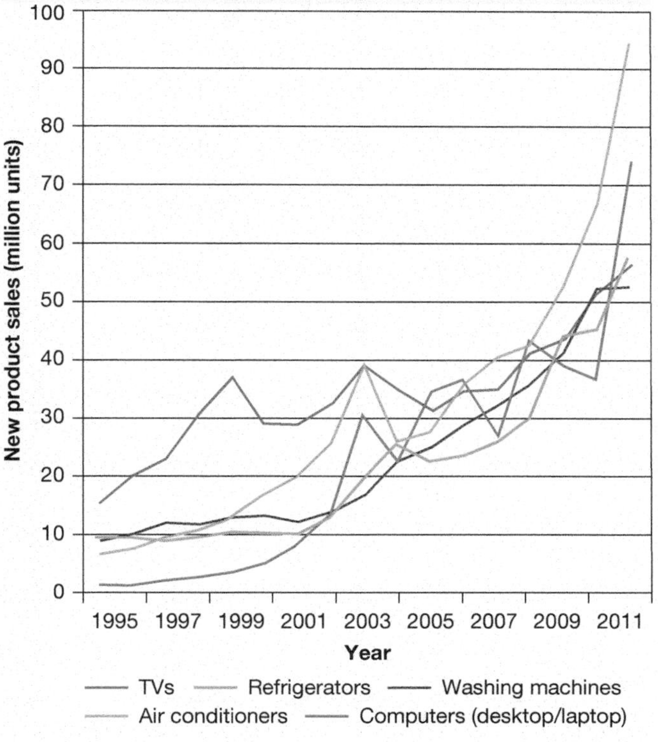

11.6.2 Production and consumption

Manufacturing accounts for about one-third of China's economic output; of this, about half is the production of electronic goods. This growing industry has resulted in an increase in

FIGURE 1 Sales of five major home appliances in China

Legend:
— TVs — Refrigerators — Washing machines
— Air conditioners — Computers (desktop/laptop)

x-axis: Year (1995–2011); y-axis: New product sales (million units)

the GDP per capita of the country. The GDP (per capita) has risen from US$ 5572 in 2011 to US$ 7590 in 2014. In 2011, 0.25 billion mobile phones were sold in China — more than any other electronic product.

11.6.3 E-waste

The importation of e-waste was banned by the Chinese Government in 2000; however, flows are still making their way into the country. The high levels of consumption of electronics globally poses a challenge for the Chinese Government as items reach the end of their life cycle or become obsolete. There are four main ways e-waste is making its way into China, even though the government has placed a ban:

• direct shipment to Chinese ports
• mixed shipments with bulk steel and copper scraps
• transit through Hong Kong
• transit through Vietnam.

11.6.4 Informal and formal collection

Formal collectors of e-waste are tax-paying businesses that recycle using environmentally sound methods. In contrast, informal collectors are self-employed, and often migrants, who travel door-to-door collecting household items for cash.

The formal collectors form a part of the national Home Appliances Old for New Rebate Program. This program was set up by the government to stimulate the purchasing of new products, and hence the economy, and to ensure the proper recycling of old appliances. Only authorised collectors such as retailers, chain stores and supermarkets are allowed to participate in this program (see figure 2).

FIGURE 2 Spatial distribution of registered formal e-waste recyclers in China

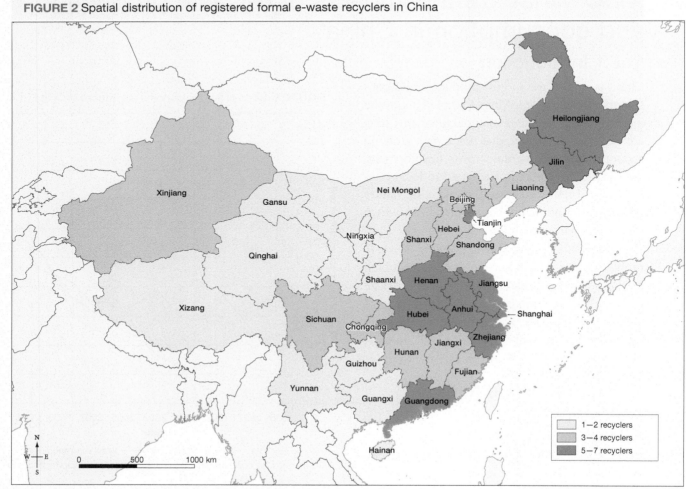

Source: E-Waste in China: A Country Report, StEP Iniative / United Nations 2012

In 2011 formal collectors collected 61.29 million home appliances, with e-waste making up about 64 per cent of this. Informal collectors work as brokers between consumers and medium-level scrap dealers. This method is convenient for households as it saves them time and also generates a small amount of money. About 20 million migrants are involved in the informal collection of household goods, many of which are e-waste. Figure 3 shows the distribution of informal sites and ports throughout China.

The formal sector is also less efficient than the informal sector as it does not have access to individual households. A similar trend is also seen in countries such as India, Nigeria and Pakistan, where the informal sector is growing and employs a lot more people.

FIGURE 3 Spatial distribution of informal e-waste recycling sites and ports through which e-waste shipments enter China

Source: E-Waste in China: A Country Report, StEP Iniative / United Nations 2012

11.6.5 Impacts of informal collection

The two largest centres for informal collection are located in the Guangdong Province and the Zhejiang Province (see figure 3). This method of collection is often labour intensive and involves contact with toxic and dangerous hazards. When workers dismantle equipment they do not use adequate protection, which leads to exposure (see figure 4).

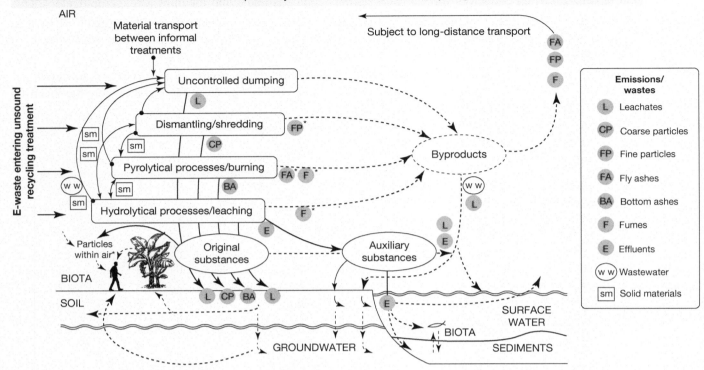

FIGURE 4 Emissions and environmental pathways from informal e-waste recycling in China

11.6.6 Law and regulations

There are five main pieces of legislation regarding the management of e-waste in China:

- The Basel Convention, passed in February 2000
- The Technical Policy on Pollution Prevention and Control, enacted in 2006
- The Ordinance on Management of Prevention and Control of Pollution from Electronic and Information Products, implemented in 2007
- Administrative Measures on Pollution Prevention of Waste and Electrical Equipment, enacted in 2008
- The Regulation on Management of the Recycling and Disposal of Waste Electrical and Electronic Equipment, implemented in January 2011.

11.6 Activities

To answer questions online and to receive **immediate feedback** and **sample responses** for every question, go to your learnON title at www.jacplus.com.au. *Note:* Question numbers may vary slightly.

Remember

1. What percentage of electronic goods is exported from China?

Explain

2. What factors have contributed to the increase in wealth in China?

Discover

3. Research one of the two main provinces that continue to participate in informal processing of e-wastes. What are the impacts on the people and the environment?

Think

4. Create a poster that gives suggestions for ways China can curb its informal collection and processing of e-wastes.

Despite this, illegal activity still occurs. Some local governments see these laws as a way of restricting progress and contrary to local interests and as a result they are not enforced. In addition, given the large number of ways e-waste is still entering China, reforms of customs controls are required.

11.7 How are e-wastes managed?

11.7.1 What is e-waste?

Litter in the schoolyard is only minor waste compared with that produced by technology and communication: e-waste. A controversial issue is how to dispose of it. The disposal and trade of e-waste is seen by developed nations as a solution and by many developing countries as a money maker. How e-waste is dealt with has serious economic, environmental and social consequences.

E-waste is any old electrical equipment, such as computers, toasters, mobile phones, iPods and televisions, that is broken, obsolete or no longer wanted. Given our technology-based lifestyle, our e-waste pile is growing at an alarming rate — faster than we know what to do with. There are five main places where e-waste usually ends up (see figure 1).

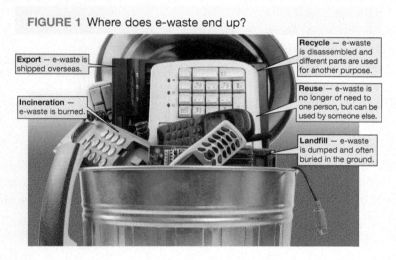

FIGURE 1 Where does e-waste end up?

Export — e-waste is shipped overseas.

Incineration — e-waste is burned.

Recycle — e-waste is disassembled and different parts are used for another purpose.

Reuse — e-waste is no longer of need to one person, but can be used by someone else.

Landfill — e-waste is dumped and often buried in the ground.

11.7.2 What are the impacts of e-waste?

Many of us do not think about where our rubbish goes. We just put it in a bin, and then it is out of sight, out of mind. However, e-waste is now a part of the global economy, and is sold, shipped and dumped worldwide.

We must think about the ramifications of such actions. Computers are made up of many toxic chemicals and metals; when dumped, these can cause significant environmental and health problems (see figure 2).

FIGURE 2 The composition of a computer

Ferrous metal	Plastic	Non-ferrous metal	Glass	Electric boards
32%	23%	18% Lead Cadmium Antimony Beryllium Mercury	15%	12% Gold Palladium Silver Platinum

Health impacts

In the slums of Delhi, computers, keyboards and monitors are pulled to pieces and then placed in acid baths or melted, to extract metals. This practice should only ever occur in proper recycling facilities, where safety precautions can take place. Consequently, there has been a rise in Delhi of **mercury poisoning** and other side effects of dealing with cadmium, plastic, PVC, barium, beryllium, and carcinogens such as carbon black and heavy metals (see figure 3).

A report by India's Department of Scientific and Industrial Research showed that e-waste being sent to India is increasing by 10 per cent each year. Slum dwellers are unsafely dis-assembling the e-waste, and toxins are damaging the health of an enormous number of people.

Exposure to mercury, even in tiny amounts, can lead to health problems and threaten child development. It damages the nervous, digestive and immune systems, and harms the lungs and kidneys. It is corrosive to the eyes and skin and to the gastrointestinal tract if swallowed. Neurological and behavioural disorders have also been observed, which can lead to insomnia, memory loss and motor dysfunction. A large number of women and children in India now expose themselves daily to mercury while working in makeshift e-waste recycling facilities in slums.

Environmental impacts

As we have seen, thousands of tonnes of e-waste are being dumped in developing countries. There, the e-waste is burned in informal recycling stations in slums. In the slums, there is nowhere to dispose of byproducts such as acid and cyanide, and they are simply dumped in the local waterways or onto the soil. This contaminates aquifers and destroys vital freshwater resources, making the land too toxic to farm.

The issue is that the e-waste can be of great value to the citizens of developing nations, such as China and India, because it is a fairly easy way to make money. For example, the circuit boards of most phones contain precious metals such as copper, gold, lead and zinc which can be extracted and sold. However, there are many poisonous substances as well, which are dangerous to people and the environment (see figure 3). For example, burning the plastic coating around copper wires gives off deadly fumes, while heavy metals dumped into local rivers creates hazardous pollution.

The burning of plastic leads to an increase in air pollution, resulting in above-standard emissions. The air is often left with the stench of acid, paint and plastic, which impairs people's quality of life. Contamination of agricultural soils may occur, affecting local farms, which are often the main form of income in developing countries.

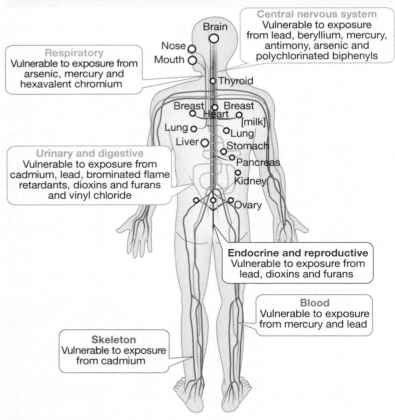

FIGURE 3 Health impacts of e-waste on waste workers and people who live near landfills or incinerators

Central nervous system
Vulnerable to exposure from lead, beryllium, mercury, antimony, arsenic and polychlorinated biphenyls

Respiratory
Vulnerable to exposure from arsenic, mercury and hexavalent chromium

Urinary and digestive
Vulnerable to exposure from cadmium, lead, brominated flame retardants, dioxins and furans and vinyl chloride

Endocrine and reproductive
Vulnerable to exposure from lead, dioxins and furans

Blood
Vulnerable to exposure from mercury and lead

Skeleton
Vulnerable to exposure from cadmium

Brain
Nose
Mouth
Thyroid
Breast
Heart
Breast
[milk]
Lung
Lung
Liver
Stomach
Pancreas
Kidney
Ovary

FIGURE 4 Two goats in a heavily polluted part of a slum in Ghana, Accra. When e-waste ends up in landfill, the toxins, carcinogens and heavy metals in it can leach into the environment. These substances then contaminate the groundwater and soil and enter the food chain.

Waste acids recovered from the burning process, which are of no value or use, are often dumped in rivers and streams, damaging the ecosystem. Often this leaves waterways black and pungent, and leads to contamination of drinking water. In Guiyu, China, the water is so toxic it can disintegrate a coin within a few hours.

11.7.3 International management of e-waste

By the start of this century, many countries could not deal with either the quantity or toxicity of the e-waste they were generating. One solution was to export it to developing nations that did not have laws in place to protect workers or the environment. It is also significantly cheaper to recycle waste in developing nations.

FIGURE 5 Who gets the trash?

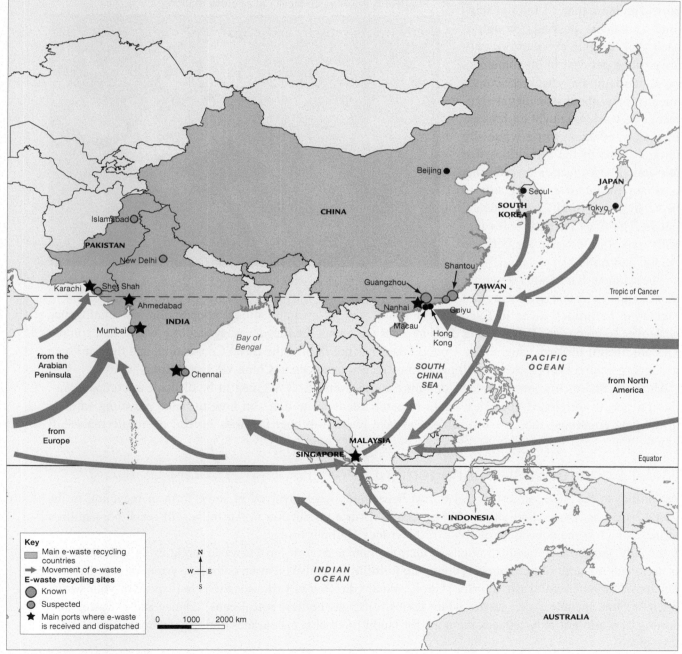

Source: Data from Greenpeace

Concerns in Europe regarding the export of e-waste led to an agreement, known as the Basel Convention, being developed to manage the movement and disposal of hazardous wastes. The convention was adopted by the European Union (EU) in 1994 and has now been accepted by more than 151 countries, including Australia. The countries that sign the convention must:

- keep the production of hazardous waste as low as possible
- make suitable disposal facilities available
- reduce and manage the international movement of hazardous waste
- ensure management of waste is controlled in an environmentally friendly way
- block and punish illegal movement of hazardous waste.

However, the EU has reluctantly admitted that its rules are only somewhat effective, as the bans on exporting e-waste are being ignored or avoided.

Rotterdam is Europe's busiest port, with more than 11 million shipping containers passing through it each year. Despite every effort, only about three per cent of all containers in Rotterdam are checked; consequently only about one illegal shipment of e-waste is caught each week. Inevitably, containers of e-waste slip through. One of the difficulties that customs officers face is the fact that it is not illegal to export goods to be re-used and sold as secondhand goods. The fines for companies illegally exporting e-waste are between A$300 and A$9000.

The United Kingdom's Environment Agency, which is in charge of regulating hazardous waste in England and Wales, has set up international cooperation and intelligence with over 40 countries. This is being done in order to combat the illegal smuggling of e-waste, and companies are fined if they are caught. The United States has not ratified the Basel Convention, and they are seen as lagging behind Europe. Almost 80 per cent of its e-waste goes to China via Hong Kong.

Until we find a cheap alternative to dumping waste overseas, it is going to be difficult to compete with this process in developed countries. More research needs to be done on how to make recycling safer in developing countries, and how to enable it to take place in proper factories instead of unsafe makeshift plants.

FIGURE 6 An e-waste facility at a refuse station

11.7.4 Local e-waste management

According to the ABS, almost all Australian households were involved in some form of recycling in the past 12 months. One report shows that by 2017, e-waste production in Australia will reach 65.4 million tonnes, which is the equivalent of about 1200 Sydney Harbour bridges.

Locally, most waste transfer stations in metropolitan areas now have e-waste facilities. A visit to the tip is no longer a matter of putting everything into **landfill**; it involves a more complex system of organisation. On arrival at your local waste transfer station, you are directed to various stations to dispose of your waste. Such stations include green waste, white goods, oils, gas bottles, petrol cans, copper and e-waste. This sorting of waste not only reduces the need for landfill, but also encourages and enables more recycling to take place.

11.7 Activities

To answer questions online and to receive **immediate feedback** and **sample responses** for every question, go to your learnON title at www.jacplus.com.au. *Note*: Question numbers may vary slightly.

Remember

1. Define the term *e-waste*. Compile a list of potential e-waste either in the classroom or in your home.
2. Why is it difficult to prevent the illegal trade in e-waste?

Explain

3. Using figure 5, explain the geographical pattern to illegal dumping of e-waste. Ensure you make specific reference to ***places*** in Asia on the map.

Discover

4. Discover the difference between ferrous and non-ferrous metals. Which is more harmful to humans and the ***environment***?
5. Research the metals listed in figure 3, and explain what makes them harmful to humans and what ***environmental*** effects they can have.

Think

6. Create a presentation on how e-waste is managed at your school. Suggest strategies for making the current management techniques more ***sustainable***.
7. Why are international agreements on e-waste difficult to manage?

learnon RESOURCES — ONLINE ONLY

Try out this interactivity: e-wasted (int-3343)

11.8 SkillBuilder: Constructing a table of data for a GIS

online only

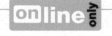

WHY ARE THERE TABLES WITHIN GIS?

AGIS, or geographical information systems, uses tables to organise and store information about points, lines and polygons (vector data). These tables have rows and columns, called fields. The GIS software links the rows in the table to the points, lines or polygons on a map.

Go online to access:

- a clear step-by-step explanation to help you master the skill
- a model of what you are aiming for
- a checklist of key aspects of the skill
- a series of questions to help you apply the skill and to check your understanding.

FIGURE 1 In a GIS, each row in a table like this may be linked to a polygon on the map.

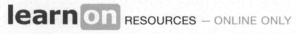

Sample	Address	No_home	No_mobiles
1	42 Jacob Street	2	4
2	27 Jacob Street	3	3
3	36 Adele Avenue	4	3
4	34 Flint Street	4	1
5	35 Flint Street	5	3
6	25 Flint Street	4	2
7	12 Jess Court	4	2
8	2 Jess Court	4	4
9	12 Flint Street	5	3
10	52 Jacob Street	6	2

learnon RESOURCES — ONLINE ONLY

Watch this eLesson: Constructing a table of data for a GIS (eles-1743)

Try out this interactivity: Constructing a table of data for a GIS (int-3361)

11.9 How does e-cycling work?

online only

Access this subtopic at **www.jacplus.com.au**

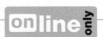

11.10 How can you reduce your consumption?

11.10.1 Act local, think global

On a global scale, entrepreneurs are tackling the waste problem society is creating. We live in a part of the world where consumption of information and communication technology is high. However, some global solutions have now been set up to ensure that individuals can easily find out online about what is recyclable, and can then undertake recycling from their own home. Society is now being encouraged to act locally, while thinking globally. This is an important concept when it comes to reducing and managing our own consumption.

11.10.2 Recycling your e-waste

In 2009, only nine per cent of the 16.8 million electronic devices discarded were recycled. About 88 per cent went to landfill, and the rest was exported as waste. Globally, 57.4 million tonnes of e-waste was recycled in 2010, and more than 75 million tonnes in 2015.

According to the ABS, when people were surveyed about their e-waste disposal, most stated that they did not recycle because they did not generate enough materials to warrant the use of e-waste services or facilities. Perhaps if everyone knew more about the hazardous nature and environmental impacts of e-waste, the idea of recycling would be embraced.

Below are some innovative ways of dealing with waste.

Ziilch

Most of us have an old computer we no longer use, or an old charger for a device we no longer have. We may even have books, furniture or clothing that could be recycled. The website Ziilch has come up with an easy solution to de-clutter, find new items and save on disposal costs. It is as easy as a click of a button (see figure 1).

This online company connects people in different places, and the platform enables you to exchange goods for free. This promotes the idea of recycling, which reduces consumption and will hopefully lead to a more sustainable future.

FIGURE 1 How Ziilch works

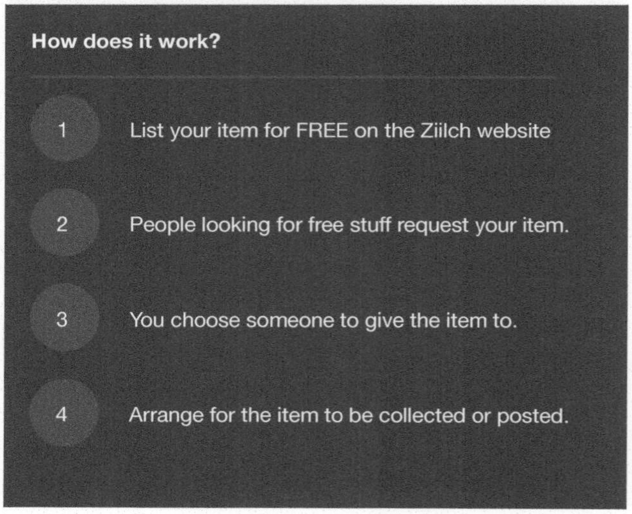
How does it work?
1
List your item for FREE on the Ziilch website
2
People looking for free stuff request your item.
3
You choose someone to give the item to.
4
Arrange for the item to be collected or posted.

MobileMuster

MobileMuster is a simple, easy way to dispose of old mobile phones, batteries and accessories. MobileMuster is voluntarily funded and managed by distributors, network carriers, service providers and manufacturers. It is a free service where you can send your mobile phone or find your nearest drop-off point, of which there are over 4500 in Australia. This ensures that mobiles are recycled correctly rather than dumped in landfill. Over 90 per cent of materials in mobile phones can be recycled (see figure 2).

By using a certified recycling program, you can ensure that your e-waste is not being illegally exported, and the environment is not being contaminated by the metals and minerals in the products.

Reduce energy consumption

It is not only important to be involved in recycling, but also to reduce your **carbon emissions**, which are another by-product of information and communication technology. Some strategies include:

- turning off your PC overnight, on weekends and during holiday periods
- ensuring printers and other equipment are switched off when not in use
- refraining from printing emails whenever possible
- removing active screensavers
- only charging your mobile phone when it is flat.

National Television and Computer Recycling Scheme

The National Television and Computer Recycling Scheme, which commenced in November 2011, provides households and small businesses with free recycling services for televisions and computers. While it is governed by the *Australian Government Product Stewardship Act of 2011*, state and territory governments retain responsibility for regulating waste.

The first drop-off points were rolled out in 2012 in metropolitan areas, and expanded into rural and remote areas in 2013 with 98 per cent of the population having reasonable access to the service. The scheme will accept all televisions, computers, and computer accessories such as keyboards and hard drives. To date, more than 130 000 tonnes of televisions and computers has been collected from more than 1800 collection points.

It is hoped that by 2021–22, 80 per cent of computers and televisions will be recycled, compared to the 17 per cent that are recycled today.

FIGURE 2 The mobile phone recycling process

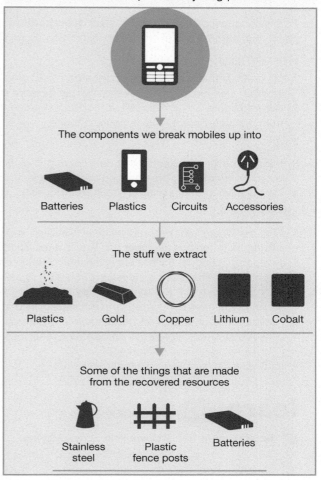

FIGURE 3 Australia's television and computer recycling program

11.10 Activities

To answer questions online and to receive **immediate feedback** and **sample responses** for every question, go to your learnON title at www.jacplus.com.au. *Note:* Question numbers may vary slightly.

Remember

1. (a) In 2009, what percentage of discarded electronic devices was recycled?
   (b) What was the main excuse that people gave for not recycling?
2. Refer to figure 2. Name three byproducts extracted from mobile phones and two products that can be made from recovered resources in mobile phones.

Think

3. Have you ever participated in a national or global recycling scheme? If so, explain what it was and what you recycled. If not, suggest what you could recycle and how to do so.
4. Create a list of things you can do in your house to reduce your energy consumption.

Discover

5. Find out where your nearest National Television and Computer Recycling Scheme drop-off point is. Describe this *place* in relation to where you live. Is it realistic for you to take your recyclables there?
6. In a group, conduct a class survey in which you interview students and teachers about their recycling habits.
   (a) Decide on the types of questions you wish to ask and how you will record responses. If you wish to do this online, use the **Survey Monkey** weblink in the Resources tab.
   (b) After you have conducted your surveys, collate and present your findings in graphic form.
   (c) Analyse your graphs and write a summary of your findings, ensuring you cover the following questions.
      • Are students more *environmentally* conscious than teachers?
      • Does age make a difference?
   (d) Prepare a recycling plan for your school.

 RESOURCES — ONLINE ONLY

🔗 **Explore more with this weblink:** Survey Monkey

11.11 SkillBuilder: Using advanced survey techniques — interviews

WHAT ARE INTERVIEWS THAT SURVEY PEOPLE'S OPINIONS?

Conducting a survey means asking questions, recording and collecting responses, and collating the number of responses.

FIGURE 1 Sample interview questions

Go online to access:

- a clear step-by-step explanation to help you master the skill
- a model of what you are aiming for
- a checklist of key aspects of the skill
- a series of questions to help you apply the skill and to check your understanding.

 RESOURCES — ONLINE ONLY

🧩 **Watch this eLesson:** Using advanced survey techniques (eles-1742)

🧩 **Try out this interactivity:** Using advanced surveying techniques (int-3360)

11.12 Review

11.12.1 Review

The Review section contains a range of different questions and activities to help you revise and recall what you have learned, especially prior to a topic test.

11.12.2 Reflect

The Reflect section provides you with an opportunity to apply and extend your learning.

Access this subtopic at **www.jacplus.com.au**

TOPIC 12
Fieldwork inquiry: What are the effects of travel in the local community?

12.1 Overview

Numerous **videos** and **interactivities** are embedded just where you need them, at the point of learning, in your learnON title at www.jacplus.com.au. They will help you to learn the content and concepts covered in this topic.

12.1.1 Scenario and your task

People travel for many reasons at the local scale — for example, they may travel to work, to shops, to visit friends and to local sporting venues. Often there are times when traffic congestion occurs, creating danger areas for motorists and pedestrians. Examples of places where such congestion occurs are schools and shopping centres. Undertaking fieldwork allows you to observe and collect original data first-hand.

Your task

Your team has been commissioned by the local council to compile a report evaluating the impacts of travel movements around a local school or traffic hotspot. You will need to collect, process and analyse suitable data and then devise a plan to better manage future traffic and pedestrian movement in the area.

12.2 Process

12.2.1 Process

- As part of a class discussion, determine a suitable location for your fieldwork study. This might be your own or a local school, or a nearby shopping centre. Talk about some of the issues related to your fieldwork site and then devise a key inquiry question — for example: What are the effects of … ? or How can we reduce the impact of … ? This will be the focus of your fieldwork. You then need to establish the following:
  - **What** sort of data and information will you need to study the travel issue at your site?
  - **How** will you collect this information?
  - **Where** would be the best locations to obtain data?
  - **When** would be the best times of the day or day(s) of the week to obtain data?
  - **How** will you record the information you are collecting?

  If you wish to collect people's views on the issue, or suggestions for improvements, you will need to plan and write suitable survey questions.

12.2.2 Collecting and recording your information and data

- As a class, plan the field trip by identifying and allocating tasks and possible sites to groups or pairs. It is often easier to share data collection. Once everything has been planned, you will need to perform your allocated tasks on the day.
- In class, invite your school principal or a member of your local council to be a guest speaker discussing your fieldwork site. They may be able to assist with background information that you may not be able to gain elsewhere. They can also provide a different perception of the effects of travel at your site. Plan a series of questions you would like to ask and be prepared to take notes that you can use in your report.
- After the field trip, it may be necessary to collate everyone's data and summarise surveys so that everyone has access to the shared information.

12.2.3 Analysing your information and data

- Look at your completed graphs and maps. What trends, patterns and relationships can you see emerging? Within your fieldwork area, are there some places that have a bigger issue with cars and pedestrians than other areas?

  Is there an interconnection between traffic congestion and time of the day, or day of the week? What have your surveys revealed? What are the major effects of travel at your fieldwork site? How do people perceive the travel issues in this place? Go back to your key inquiry question. To what extent have you been able to answer it? Write your observations up as a fieldwork report using subheadings such as:
  - Background and key inquiry question
  - Conducting the fieldwork [planning and collecting data]
  - Findings [results of data analysis].
- Download the report template from the Resources tab to help you complete this project. Use the report template to create your report.

12.2.4 Communicating your findings

- Now that you have identified a traffic problem and collected and analysed data, it is time to try to solve it. Your completed map and supporting data will form part of your management plan for the future. What have been the main issues that have emerged from your fieldwork research? How can you best manage these issues? Using your base map, create an overlay or annotated map to show possible options for reducing the traffic problem. You will need to support each proposal with data that you have gained from your fieldwork. Possible ideas could include:
  - changing parking restrictions
  - staggering times of drop-off and pick-up
  - introducing traffic wardens to guide traffic
  - creating a one-way system.

Your teacher may arrange for your completed report to be presented to your school or local council. Considering your audience, what is the best way to present your findings? You might like to produce a PowerPoint presentation or an annotated visual display.

12.3 Review

12.3.1 Reflecting on your work

- Think back over how well organised and prepared you were for the fieldwork, the data you collected and how you processed the data for your report. Download and complete the reflection template from the Resources tab.

GLOSSARY

active consumerism: a movement that is opposed to the endless purchase of material possessions and the pursuit of economic goals at the expense of society or the environment

agribusiness: business set up to support, process and distribute agricultural products

agroforestry: the use of trees and shrubs on farms for profit or conservation; the management of trees for forest products

aquaculture: the farming of aquatic plants and aquatic animals such as fish, crustaceans and molluscs

aquaponics: a sustainable food production system in which waste produced by fish or other aquatic animals supplies the nutrients for plants, which in turn purify the water

aquifer: a body of permeable rock below the Earth's surface, which contains water, known as groundwater

arable: describes land that can be used for growing crops

barter: to trade goods in return for other goods or services rather than money

biodiesel: a vegetable oil or animal fat-based diesel fuel

biodiversity: the variety of plant and animal life within an area

biofuel: fuel that comes from renewable sources

biophysical environment: the natural environment, made up of the Earth's four spheres — the atmosphere, biosphere, lithosphere and hydrosphere

black market: any illegal trade in officially controlled or scarce goods

canal housing estate: a housing estate built upon a system of waterways, often as the result of draining wetland areas. All properties have water access.

carbon emissions: carbon dioxide that is released into the atmosphere by natural processes or by burning gas, coal or oil contributing to climate change

cash crop: a crop grown to be sold so that a profit can be made, as opposed to a subsistence crop, which is for the farmer's own consumption

clearfelling: the removal of all trees in an area

coral polyps: a tube-shaped marine animal that lives in a colony and produces a stony skeleton. Polyps are the living part of a coral reef.

crop rotation: a procedure that involves the rotation of crops, so that no bed or plot sees the same crop in successive seasons

deforestation: clearing forests to make way for housing or agricultural development

degradation: deterioration in the quality of land and water resources caused by excessive exploitation

de-manufacture: to disassemble devices into their original components for recycling

desertification: the transformation of arable land into desert, which can result from climate change or from human practices such as deforestation and overgrazing

developed: describes countries with a highly developed industrial sector, a high standard of living, and a large proportion of people living in urban areas

developing countries: nations with a low living standard, undeveloped industrial base and low human development index relative to other countries

digital divide: a type of inequality between groups in their access to and knowledge of information and communication technology

disability: a functional limitation in an individual, caused by physical, mental or sensory impairment

discretionary item: an item that is bought out of choice, according to one's judgement

ecotourism: tourism that interprets the natural and cultural environment for visitors, and manages the environment in a way that is ecologically sustainable

edible: fit to be eaten as food; eatable

endemic: describes species that occur naturally in only one region

environmental refugees: people who are forced to flee their home region due to environmental changes (such as drought, desertification, sea-level rise or monsoons) that affect their wellbeing or livelihood

erosion: the wearing down of rocks and soils on the Earth's surface by the action of water, ice, wind, waves, glaciers and other processes

e-Stewards certification: an e-waste standard that bans exporting of hazardous electronic wastes to developing countries and outlaws the use of prisoners searching local dumps and landfills for e-waste

ethnicity: cultural factors such as nationality, culture, ancestry, language and beliefs

e-waste: any old electrical equipment such as computers, toasters, mobile phones and iPods that no longer works or is no longer required

expenditure: the amount of money spent

extensive farm: farm that extends over a large area and requires only small inputs of labour, capital, fertiliser and pesticides

extremism: extreme political or religious views or extreme actions taken on the basis of those views

factory farming: the raising of livestock in confinement, in large numbers, for profit

famine: a drastic, widespread food shortage

genetically modified: describes seeds, crops or foods whose DNA has been altered by genetic engineering techniques

globalisation: the process that enables markets and companies to operate internationally, and world views, products and ideas to be freely exchanged

graffiti: the marking of another person's property without permission; it can include tags, stencils and murals

Green Revolution: a significant increase in agricultural productivity resulting from the introduction of high-yield varieties of grains, the use of pesticides and improved management

greenhouse gases: any of the gases that absorb solar radiation and are responsible for the greenhouse effect. The three main gases are water vapour, carbon dioxide and methane.

gross domestic product (GDP): the value of all the goods and services produced within a country in a given period of time. It is often used as an indicator of a country's wealth.

groundwater: water that exists in pores and spaces in the Earth's rock layers, usually from rainfall slowly filtering through over a long period of time

halal: describes food that is prepared under Islamic dietary guidelines

hard sport tourism: tourism in which someone travels to either actively participate in or watch a competitive sport as the main reason for their travel

horticulture: the practice of growing fruit and vegetables

household final consumption per person: the value of each person's goods and services expenditure in a country

humanitarian aid: assistance provided in response to human crises caused by natural or man-made disasters, in order to save lives, relieve suffering and maintain human dignity

humanitarian principles: the principles governing our response to natural disasters, such as tsunamis and earthquakes, or human-induced disasters, such as armed conflicts — the main aim being to save lives and alleviate suffering

humus: an organic substance in the soil that is formed by the decomposition of leaves and other plant and animal material

hybrid: plant or animal bred from two or more different species, sub-species, breeds or varieties, usually to attain the best features of the different stocks

hydroponic: describes a method of growing plants using mineral nutrients, in water, without soil

income diversity: income that comes from many sources

indicator: something that provides a pointer, especially to a trend

infrastructure: the facilities, services and installations needed for a society to function, such as transportation and communications systems, water pipes and power lines

innovation: new and original improvement to something, such as a piece of technology or a variety of plant or seed

intensive farm: farm that requires a lot of inputs, such as labour, capital, fertiliser and pesticides

irrigation: the supply of water by artificial means to agricultural areas where there is a shortage

jatropha: any plant of the genus *Jatropha*, but especially *Jatropha curcas*, which is used as a biofuel

kenaf: a plant in the hibiscus family that has long fibres; useful for making paper, rope and coarse cloth

landfill: a method of solid-waste disposal in which refuse is buried between layers of soil

latitude: the angular distance north or south from the equator of a point on the Earth's surface

leaching: the process by which water runs through soil, dissolving minerals and carrying them into the subsoil

leeward: describes the area behind a mountain range, away from the moist prevailing winds

lithium ion battery: type of rechargeable battery that is 'energy dense' and does not lose much charge when not in use

logging: large-scale cutting down, processing and removal of trees from an area

low-income country: a country that has a gross national income (GNI) per capita of $1025 or less, has not reached its final stage of development and is eligible for international development assistance (IDA)

mallee: vegetation areas characterised by small, multi-trunked eucalypts found in the semi-arid areas of southern Australia

malnourished: describes someone who is not getting the right amount of the vitamins, minerals and other nutrients to maintain healthy tissues and organ function

marginal land: describes agricultural land that is on the margin of cultivated zones and is at the lower limits of being arable

Masai: an ethnic group of semi-nomadic people living in Kenya and Tanzania

mature-aged: describes individuals aged over 55

median age: the age that is in the middle of a population's age range, dividing a population into two numerically equal groups

mercury poisoning: a toxic condition caused by the ingestion or inhalation of mercury or a mercury compound. It has various symptoms, including vomiting, nausea, insomnia and fevers.

middle-income country: a country that has a gross national income (GNI) per capita of $1026 to $12 479 and has not reached its final stage of development

mobility: the ability to move or be moved freely and easily

monoculture: the cultivation of a single crop on a farm or in a region or country

national park: a park or reserve set aside for conservation purposes

national security: the protection of a nation's citizens, natural resources, economy, money, energy, environment, military, government and energy

new moon: the phase of the moon when it is closest to the sun and is not normally visible

nomadic: describes a group of people who have no fixed home and move from place to place according to the seasons, in search of food, water and grazing land

non-government organisations: non-profit group run by people (often volunteers) who have a common interest and perform a variety of humanitarian tasks at a local, national or international level

offshored: to relocate part of a company's processes or services overseas in order to decrease costs

old-growth forests: natural forests that have developed over a long period of time, generally at least 120 years, and have had minimal unnatural disturbance such as logging or clearing

organic matter: decomposing remains of plant or animal matter

per capita: per person (literally 'by head')

perception: the process by which people translate sensory impressions into a view of the world around them

plantations: an area in which trees or other large crops have been planted for commercial purposes

pneumatophores: exposed root system of mangroves, which enable them to take in air when the tide is in

potable: drinkable; safe to drink

prairie: native grasslands of North America

precipitation: the forms in which moisture is returned to the Earth from the sky, most commonly in the form of rain, hail, sleet and snow

primary industry: industry involved in the gathering of natural resources, such as iron ore and timber, or activities such as farming and fishing

pulp: the fibrous material extracted from wood or other plant material to be used for making paper

rain shadow: the dry area on the leeward side of a mountain range

Ramsar site: a wetland of international importance, as defined by the Ramsar Convention — an intergovernmental treaty on the protection and sustainable use of wetlands

robotic technology: the branch of technology that deals with the design, construction, operation and application of robots and computer systems

salinity: the presence of salt on the surface of the land, in soil or rocks, or dissolved in rivers and groundwater

seasonal crops: crops that are harvested in a certain season of the year, rather than all year round

smuggling: importing or exporting goods secretly or illegally

social justice: the aim to create a society that is based on equality, that values human rights and that recognises the dignity of every human being

soft sport tourism: tourism in which someone participates in recreational and leisure activities, such as skiing, fishing and hiking, as an incidental part of their travel

stereotypes: widely held but oversimplified idea of a type of person or thing

street art: artistic work done with permission from both the person who owns the property on which the work is being done and the local council

subsistence: describes farming that provides food only for the needs of the farmer's family, leaving little or none to sell

sustainable: describes the use by people of the Earth's environmental resources at a rate such that the capacity for renewal is ensured

totem: an animal, plant, landscape feature or weather pattern that identifies an individual's connection to the land

trade barriers: government-imposed restriction (in the form of tariffs, quotas and subsidies) on the free international exchange of goods or services

trading partners: a participant, organisation or government body in a continuing trade relationship

treaty: a formal agreement between two or more independent states or nations, and usually involving a signed document

treeline: the edge of the areas in which trees are able to grow

tundra: the area lying beyond the treeline in polar or alpine regions

undernourished: describes someone who is not getting enough calories in their diet; that is, not enough to eat

undulating: describes an area with gentle hills

urbanisation: the growth and spread of cities

value adding: processing a material or product and thereby increasing its market value

water stress: situation that occurs when water demand exceeds the amount available or when poor quality restricts its use

waterlogging: saturation of the soil with groundwater so that it hinders plant growth

watertable: the surface of the groundwater, below which all pores in the soils and rock layers are saturated with water

Web 1.0: a read-only version of the web. It enabled users only to search and read information.

Web 2.0: version of the web that allows the sharing of online information and ideas that anyone has created. It is the second generation of the World Wide Web that enables collaboration and exchanging of information online.

Western-style diet: eating pattern common in developed countries, with high amounts of red meat, sugar, high-fat foods, refined grains, dairy products, high-sugar drinks and processed foods

windward: describes the side of the mountain that faces the prevailing winds

winter solstice: the shortest day of the year, when the sun reaches its lowest point in relation to the equator

yield gap: the gap between a certain crop's average yield and its maximum potential yield

INDEX

A

Aboriginal peoples *see* Indigenous Australian peoples
accessibility, for people with disabilities 153–6
active consumerism 121
active travel 146–8
activity hubs 152
Adelaide, revitalisation of CBD 142
Africa, land grabs 97–8
agribusiness 55
agricultural innovations 39
agricultural technician 2
agriculture
  global warming and 87
  modification of biomes 53–4, 85
  potential golden era in Australia 126
  types in Australia 55–6
aid *see* food aid; foreign aid
air movements, climate and 32
Airbus A380, production of 184
altitude, influence on climate 31
Anangu Pitjantjatjara Yankunytjatjara (APY) lands program 120
aquaculture 74, 77–8
aquaponics 127
aquatic biomes, types 19
aquifers 100
arable land
  access to 93
  per capita 37
Asia
  car industry 195–6
  changing diets 125
  rice production 62–5
Asia–Pacific Economic Cooperation (APEC) forum 188
atmosphere 68
Australian Aid 203–5
Australian Government Product Stewardship Act 2011 (Cwlth) 229
Australian Trade Commission (Austrade) 188
automotive trade 195–7

B

Ballarat, accessibility 155
bartering 191
Basel Convention 226
Bedouin people 44
bike paths 147

biodiesel 197
biodiversity
  in Australia 86
  loss of 29, 69, 84–6
  megadiverse countries 86
biofuel crops 41, 95–8
biomes
  in Australia 28–30
  climatic influences 30–2
  differences 30–4
  major types and locations 17–19
  modification for agriculture 53–4, 85
  role of soil 32–4
biophysical world
  Earth's spheres 68
  impact of food production 69
biosphere 68
Bogong moths 48
Borås, Sweden 154
Brisbane, laneways 142

C

car industry
  in Asia 195–6
  in United States 196–7
carbon emissions 229
careers in Geography 1–2
change, as geographic concept 7
China
  e-waste 220
  foreign-owned enterprises 199
  production and consumption 219–20
  transport system 177–8
Chinese New Year, impact of 176–7
cities
  accessibility 153
  encroachment onto farmland 97
clearfelling 30
climate
  major influences on 31–2
  modification for agriculture 53–4
  soil and 32
climate change, food security and 103–5
clothing industry, globalisation and 198
coastal wetlands
  as biomes 22
  importance 22–3
communication technology 209
computers, composition 223
coral polyps 25

coral reefs
  benefits of 27
  formation 25–6
  threats to 27
crop rotation 64
cultural tourism 176–8
cumulative line graphs 194–5

D

deforestation 17, 71
degradation of resources 69
democracy, and internet access 216
Department of Foreign Affairs and Trade (DFAT) 188, 203
desertification 81, 95
deserts
  in Australia 30
  as biomes 18
developed countries 164
developing countries 191
dietary changes, and food supply 124–6
digital divide
  bridging gap 216
  nature of 214–16
disabilities, people with, equal access for 154
discretionary items 121
divergence graphs 171

E

e-waste
  in China 219–22
  environmental impacts 224–5
  health impacts 223–4
  impacts 223–5
  international management 225–6
  local management 226
  meaning of 223
  recycling of 220–2, 228–9
economic equality, and internet access 216
economic growth, and internet access 216
ecosystems *see* biomes
ecotourism 174–5
education-related travel services 189
eel farming 49
endemic species 86
energy consumption, reducing 229
environment, as geographic concept 8

environmental managers 2
environmental refugees 104
erosion 80
ethnicity 136
Eugene, Oregon 150
European Union (EU), e-waste
    management 226
exports 189
extensive farms 55
extremism 203

F
fair trade 191, 200–2
Fairtrade certification 122–3, 202
Fairtrade International 201
famines, development 95
farming yields, factors affecting 111
farms, types in Australia 57–8
fieldwork
    communicating findings 234
    data and information
      analysis 233
    data and information collection and
      recording 233
    process 233
    reflection 234
financial crisis 97
Financial Deepening Challenge
    Fund (FDCF) 217
fish
    as food staple 41–2, 74–5
    overfishing 75
fish farming 74, 77–8, 127
flow maps 206
food aid
    in Australia 118–20
    cash vouchers 117
    major donor countries 116
    need and delivery 115–16, 118–20
    recipients 116
    school feeding programs 117
food crisis 97
food insecurity
    among contemporary Indigenous
      peoples 48, 120
    causes of 93
    changing distribution of world
      hunger 92
    countries at most risk 92
    impact 93
food problems 37
food production
    in Australia 55–8
    environment and 50–1
    future changes 111–12
    Green Revolution 50, 52
    impact on biomes 68–9

increases in 38–9, 50–1
solutions to problems 112–14
strategies for improving 112
trade and economic factors 51
water use 84, 100–2
food security
    across the world 89–91
    climate change and 103–5
    meaning of 89
    risks of land grabs 97–8
    in traditional Indigenous
      society 46–9
    see also food insecurity
Food Security Risk Index 91, 92
food staples
    global protein demand 74
    major food types 40–2
food supply, dietary changes
    and 124–6
food trade 191–3
foreign aid, Australia as
    donor 203–5
forests
    as biomes 17
    impact of clearing 72
    importance 70
    reasons for clearing 71
freshwater biomes, types
    and locations 19
fuel crisis 97

G
genetically modified (GM)
    foods 112
geographic concepts
    applying 11–12
    change 7
    environment 8
    interconnection 6
    place 5
    scale 4
    space 10
    SPICESS 3
    sustainability 9
geographical cartoons 79
geospatial skills 1
GIS (geographical information
    systems) 1, 73
    tables of data 227
global warming, causes 87
globalisation, and industrial
    landscape 198–9
good and services
    consumption of 185
    trade in 184–5
graffiti 142
    grasslands

in Australia 29
    as biomes 15
    characteristics 18, 20
    importance 21
grazing 57
Great Barrier Reef 5, 27
Green Revolution 50, 52
greenhouse gases 70
greenhouses 113–14
Griffith, NSW 59–60
gross domestic production
    (GDP) 160
groundwater 68
Gunditjmara people 49

H
halal 193
horticulture 54
household final consumption
    per person 186
Huli people 45
Human Development Index 204
humanitarian principles 203
humus 80
hunger
    impacts 108–9
    problem of 108
    see also food insecurity
hunters and gatherers 44, 46
hybrids 52
hydroponics 127
hydrosphere 68, 69

I
illegal wildlife trade 206
imports 189–90
Indigenous Australian peoples
    connections to places 141
    contemporary food insecurity 48
    perception of land 139
    relationship with land 139
    traditional food security 46–9
    use of fire 47–8
    'Welcome to Country'
      tradition 141
industry, levels of 185
information and communication
    technology (ICT)
    expenditure on 212–14
    uses and impact 210–11
infrastructure 144, 162
intensive farming 55, 57–8
interconnection, as geographic
    concept 6
international students 189
internet access
    across the world 212

impact of unequal access 214–16
internet use 209–10, 213
interviews, using 230
irrigation
    in Griffith, NSW 59–60
    impacts on environment 83
    purpose 82
    rice production and 63, 64
    and water use 100–2
isoline maps 153

J
jatropha 97, 98

K
Kenya, M-Pesa (mobile
    money) 217–18
Kolkata wetland system 128

L
land
    competition for 96–9
    importance to people 139
    Indigenous Australian peoples'
        perception of 139
    marginal land 97, 140
    ownership of 139
land degradation
    causes 80
    impacts 81, 96
    nature of 80
    types 96
land grabs 97–8
landfill 226
landscape architects 2
landscapes, modification for
    agriculture 54
laneways
    in Melbourne 142–3
    revival around Australia 142–3
latitude 31–2
    climate and 31–2
    soil and 33
leeward 31
line graphs 194–5
lithium ion batteries 197
lithosphere 68
live animal trade 193
livestock farming 57
logging 29
low-income countries 212, 214–15
Lunar New Year 177

M
M-Pesa (mobile money),
    Kenya 217–18
maize, as food staple 41

mallee 57
malnourishment 93, 108
maps, proportional circles on 123–4
marginal land 97, 140
marine biomes, types and
        locations 19
Masai people, Kenya 162, 218
mature-aged people 164
Meals on Wheels 119, 120
meat consumption 125
median age, Australia 119
medical tourism 162–3
Médicens sans Frontières (Doctors
        without Borders) 110
Melbourne
    20-minute neighbourhoods 149–50
    accessibility for all 155–6
    bike paths 147
    laneways 142–3
    liveability 149
    public transport 145–6
    walkability 149–51
mental maps 136–7
mercury poisoning 223
meteorologists 2
middle-income countries 214
mixed farms 57
mobile phones
    disposal 229
    in Kenya 218
    use 210, 212
MobileMuster 228–9
mobility 155
monoculture 64, 80, 99
mountains, influence on climate 31
multiple line graphs 194–5

N
national parks 162
national security 203
National Television and Computer
        Recycling Scheme 229
neighbourhoods, accessibility 151–2
new moon 177
nomadic herders 44
nomadic peoples 44
non-government organisations
        (NGOs), and fair trade 202

O
ocean currents, climate and 32
offshoring of production 198–9
old-growth forest, in Australia 29–30
Olympic Games, impact of 180–1
online shopping 198
organic matter 32
Oxfam 202

P
paper production 71
park rangers 2
place, as geographic concept 5
places
    changes to 142–3
    Indigenous connections to 141
    mapping 137
    perceptions of 136
plantation farming 58, 71
Plumpy'nut 109–10
pneumatophores 23
population, Australia 2016 119
population forecasts
    global; 2010–50 109
    population pyramid,
        Australia 2050 119
Port Augusta, experimental
        greenhouse 113–14
potable water 102
prairies 20
precipitation 31
primary industry 185
proportional circles on maps 123
public transport 144–6

Q
quaternary industry 185
quinoa 125

R
rain shadows 31
Ramsar sites 128
ready-to-use therapeutic food
        (RUFT) 109
recycling, e-waste 220–1, 228–9
resource degradation 69
rice paddies 62
rice production
    in Asia 62–3
    in Australia 64–5
    biotechnology and 63
    climate and topography 62, 64
    environmental issues 64, 65
    importance 62
    irrigation 63, 64
    pests and diseases 63, 64–5
    technology use 63, 65
    world rice production 62
rivers, in Australia 28
robotic technology 197

S
salinity 80, 83, 95
San people (Kalahari Bushmen) 44
satellite images, interpreting 100
savanna *see* grasslands

scale, as geographic concept   3, 4
scattergrams, constructing   118
seagrass meadows, in Australia   29
seasonal crops   122
secondary industry   185
SecondBite   119
sharks   76–7
shifting agriculture   45
skills   1
social mobility, and internet
      access   216
soil
   differences   32–3
   role in biomes   32–4
soil profiles, layers   32
soils, modification   54
space, as geographic concept 10
spatial relationships, in thematic
      maps   16
spatial technologies   1
species extinctions   84
SPICESS   3
sports tourism   179–81
stereotypes   136
street art   142
subsistence agriculture   43, 71
surveyors   2
surveys
   creating   170
   interviews   230
sustainability, as geographic
      concept   9
sustainable agriculture   38
sweatshops   199

T
20-minute neighbourhoods   149–50,
      152
tables of data for GIS   227
ternary graphs, constucting   50
tertiary industry   185
Thanksgiving, impact of   176
thematic maps, spatial
      relationships in   16
topographic maps
   patterns and correlations   61
   transects   25
topological maps, interpreting   138
Torres Strait Islander peoples *see*
      Indigenous Australian peoples
totems   139
tourism
   accommodation   165
   by Australians   166
   cultural tourism   176–9
   ecotourism   174–5
   future of   161–2

   importance   159–60
   increase in   161, 164
   medical tourism   162–3
   negative impacts   172
   positive impacts   173
   to Australia   167
   trends   161
tourist destinations
   Australia   167–70
   sport and   179–82
   top 10 countries   165
   types   160
tourists
   top spenders   165
   types   161
trade
   automotive trade   195–7
   in animals   193
   in food   191–2
   in goods and services   185–6
   impacts of   120
   organisations   187–8
   social justice and   200
   types in Australia   189–91
trade barriers   187
trading partners   188
transects, constructing and
      describing   25
'triple F' crisis   97
tundra, as biomes   18

U
undernourishment   93
undulating land   54
United Nations Food and Agriculture
      Organization (UN FAO)   38,
      42, 111
United Nations Sustainable
      Development Goals
      (SDGs)   204
United States, car industry   196–7
urban farming   126–8
urbanisation   21

V
value adding   185

W
Wadi As Sirhan Basin, Saudi
      Arabia   12
walkability   149–51
Walkability Index   151
water availability, projections   102
water insecurity   101
water quality   102
water scarcity, causes   100–1
water stress   101

waterlogging   83
watertable   83
web 1.0   210
web 2.0   210
Western-style diet   125
wetlands
   in Australia   29
   Kolkata wetland system   128
wheat, as food staple   41–2
wheat farms   57
windward   31
winter solstice   177
World Fair Trade Organization   201
World Food Programme
      (WFP)   115–16
World Trade Organization
      (WHO)   187
World Vision   202
World Wide Web   210

Y
yield gap   112, 113
Yunnan Province, China   62

Z
Ziilch   228